HSC Year 12
MATHEMATICS EXTENSION 2

JIM GREEN I JANET HUNTER
SERIES EDITOR: ROBERT YEN

A+

· 2020 UPDATED SYLLABUS · 2020 UPDATED SYLLABUS · 2020 UPDATED SYLLABUS

+ topic exams of HSC-style questions
+ practice HSC and mini-HSC exams
+ worked solutions with expert comments
+ HSC exam topic grids (2011–2020)

PRACTICE
EXAMS

A+ HSC Mathematics Extension 2 Practice Exams
1st Edition
Jim Green
Janet Hunter
ISBN 9780170459273

Publishers: Robert Yen, Kirstie Irwin
Project editor: Tanya Smith
Cover design: Nikita Bansal
Text design: Alba Design
Project designer: Nikita Bansal
Permissions researcher: Corrina Gilbert
Production controller: Karen Young
Typeset by: Nikki M Group Pty Ltd

Any URLs contained in this publication were checked for currency during the production process. Note, however, that the publisher cannot vouch for the ongoing currency of URLs.

NSW Education Standards Authority (NESA): Higher School Certificate Examination Mathematics Extension 2 (2001, 2016, 2017, 2018, 2020) © NSW Education Standards Authority for and on behalf of the Crown in right of the State of New South Wales.

© 2021 Cengage Learning Australia Pty Limited

For product information and technology assistance,
in Australia call **1300 790 853**;
in New Zealand call **0800 449 725**

For permission to use material from this text or product, please email
aust.permissions@cengage.com

ISBN 978 0 17 045927 3

Cengage Learning Australia
Level 7, 80 Dorcas Street
South Melbourne, Victoria Australia 3205

Cengage Learning New Zealand
Unit 4B Rosedale Office Park
331 Rosedale Road, Albany, North Shore 0632, NZ

For learning solutions, visit **cengage.com.au**

Printed in China by 1010 Printing International Limited.
1 2 3 4 5 6 7 25 24 23 22 21

ABOUT THIS BOOK

Introducing *A+ HSC Year 12 Mathematics*, a new series of study guides designed to help students revise the topics of the new HSC maths courses and achieve success in their exams. *A+* is published by Cengage, the educational publisher of *Maths in Focus* and *New Century Maths*.

For each HSC maths course, Cengage has developed a STUDY NOTES book and a PRACTICE EXAMS book. These study guides have been written by experienced teachers who have taught the new courses, some of whom are involved in HSC exam marking and writing. This is the first study guide series to be published after the first HSC exams of the new courses in 2020, so it incorporates the latest changes to the syllabus and exam format.

This book, *A+ HSC Year 12 Mathematics Extension 2 Practice Exams*, contains topic exams and practice HSC exams, both written and formatted in the style of the HSC exams, with spaces for students to write answers. Worked solutions are provided along with the authors' expert comments and advice, including how each exam question is marked. An HSC exam topic grid (2011–2020) guides students to where and how each topic has been tested in past HSC exams.

Mathematics Extension 2 topics

1. Proof
2. 3D vectors
3. Complex numbers
4. Further integration
5. Mechanics

This book contains:

- 5 topic exams: 1-hour mini-HSC exams on each topic + worked solutions
- 2 practice mini-HSC exams: 1-hour exams + worked solutions
- 2 practice HSC exams: full (3-hour) exams + worked solutions
- HSC exam reference sheet of formulas
- bonus: worked solutions to the 2020 HSC exam.

The companion A+ STUDY NOTES book is written by the same authors, Jim Green and Janet Hunter (see p.ix), who also wrote the popular textbook *Maths in Focus 12 Mathematics Extension 2*. It contains topic summaries and graded practice questions, grouped into the same 5 broad topics, including for each topic a concept map, glossary and HSC exam topic grid.

Both *A+* books can be used for revision after a topic has been learned, as well as for preparation for the trial and HSC exams. Before you begin any questions, make sure you have a thorough understanding of the topic you will be undertaking.

iv

CONTENTS

CHAPTER 1

PROOF

CHAPTER 2

3D VECTORS

CHAPTER 3

COMPLEX NUMBERS

CHAPTER 4

FURTHER INTEGRATION

CHAPTER 5

MECHANICS

9780170459273

YEAR 12 COURSE OVERVIEW

PROOF

The language and methods of proof

- If-then (implication)
- Converse
- Equivalence
- Negation
- Contrapositive
- Sets of numbers
- Proof by contradiction
- Proof by counterexample

Proofs involving numbers and inequalities

- Properties of inequalities
- 'Consider the difference'
- The arithmetic mean-geometric mean inequality
- The triangle inequality

Mathematical induction

- Series and sigma notation
- Divisibility
- Inequalities
- Calculus, probability and geometry
- Recursive formulas

3D VECTORS

Operations with vectors

- 3D vectors
- Unit vectors
- Addition and subtraction
- Multiplication by a scalar
- Magnitude of a vector
- Scalar (dot) product

Vector equations of curves

- 3D space and coordinates
- Parametric equations of curves
- Equation of a sphere

Vector equations of lines

- Vector equation of a line $\underline{r} = \underline{a} + \lambda\underline{b}$
- Testing whether a point lies on a line
- Parallel and perpendicular lines

Geometrical proofs in 2D and 3D

9780170459273

COMPLEX NUMBERS

Complex numbers

- Cartesian form $z = a + ib$
- $\text{Re}(z)$, $\text{Im}(z)$
- Adding, subtracting, multiplying, dividing
- Complex conjugate, $\overline{z}$
- Realising the denominator
- Reciprocal
- Square root of a complex number

The complex plane and polar form

- The complex plane, Argand diagram
- Polar (modulus-argument) form
- Properties of modulus and argument
- Multiplying and dividing complex numbers
- Powers of complex numbers

Euler's formula and exponential form

- Euler's formula $e^{i\theta} = \cos\theta + i\sin\theta$
- Exponential form $z = re^{i\theta}$
- Exponential, Cartesian, polar forms
- Powers of complex numbers in exponential form

De Moivre's theorem and solving equations

- De Moivre's theorem
- Solving quadratic equations
- Quadratic equations with complex coefficients
- Polynomial equations and conjugate roots

Complex numbers as vectors

- Adding and subtracting complex numbers
- Multiplying complex numbers
- Conjugates, multiplying by a scalar, multiplying by i

Roots, curves and regions

- Roots of unity: location on the unit circle
- Roots of a complex number
- Graphing lines and curves in the complex plane: equations involving modulus and argument
- Graphing regions in the complex plane: inequalities

FURTHER INTEGRATION

Integration by substitution

- Let $u = \ldots$
- The substitution may not be given
- Change limits of definite integrals
- Trigonometric substitutions, including t-formulas

Rational functions and partial fractions

- Quadratic denominators that require completing the square
- Linear and quadratic denominators
- Partial fractions: equating coefficients vs substitution
- Often involve logarithmic or inverse trigonometric functions

Integration by parts

- $\int uv'\,dx = uv - \int vu'\,dx$
- Choosing u and v'

Integration by recurrence relations

- Integrals I_n, I_{n-1} and I_0
- Usually involves integration by parts

MECHANICS

Simple harmonic motion

- Velocity and acceleration as functions of displacement
- Equations and graphs for simple harmonic motion
- Acceleration, velocity and displacement
- Amplitude, period, phase shift, centre
- Properties of simple harmonic motion: maximum and zero velocity and acceleration

Resisted motion

- Resisted horizontal motion
- $R = kv$ or kv^2
- $v = f(t)$ and $v = f(x)$
- Resisted vertical motion under gravity and other forces
- Terminal velocity

Modelling motion

- Newton's laws of motion: $F = m\ddot{x}$
- Forces and resolving forces
- Force diagrams
- Coefficient of friction

Projectiles and resisted motion

- Equation of the path of a projectile
- Projectile motion under gravity and other forces

SYLLABUS REFERENCE GRID

Topic and subtopics	A+ HSC Year 12 Mathematics Extension 2 Practice Exams chapter
PROOF	
MEX-P1 The nature of proof	1 Proof
MEX-P2 Further proof by mathematical induction	1 Proof
VECTORS	

MEX-V1 Further work with vectors
    V1.1 Introduction to three-dimensional vectors
    V1.2 Further operations with three-dimensional vectors
    V1.3 Vectors and vector equations of lines
	2 3D vectors

COMPLEX NUMBERS	

MEX-N1 Introduction to complex numbers
    N1.1 Arithmetic of complex numbers
    N1.2 Geometric representation of a complex number
    N1.3 Other representations of complex numbers
	3 Complex numbers

MEX-N2 Using complex numbers
    N2.1 Solving equations with complex numbers
    N2.2 Geometrical implications of complex numbers
	3 Complex numbers

CALCULUS	
MEX-C1 Further integration	4 Further integration
MECHANICS	

MEX-M1 Applications of calculus to mechanics
    M1.1 Simple harmonic motion
    M1.2 Modelling motion without resistance
    M1.3 Resisted motion
    M1.4 Projectiles and resisted motion
	5 Mechanics

ABOUT THE AUTHORS

Jim Green was Head of Mathematics at Trinity Catholic College, Lismore, where he spent most of his teaching career of over 35 years. He has written HSC examinations and syllabus drafts, composed questions for the Australian Mathematics Competition and recently co-authored *Maths in Focus 12 Mathematics Extension 2*.

Janet Hunter is Head of Mathematics at Ascham School, Edgecliff, where she has spent most of her teaching career of over 30 years. She has been a senior HSC examiner and judge, an HSC Advice Line adviser, and recently co-authored *Maths in Focus 12 Mathematics Extension 2*.

A+ DIGITAL FLASHCARDS

Revise key terms and concepts online with the A+ Flashcards. Each topic for this course has a deck of digital flashcards you can use to test your understanding and recall. Just scan the QR code or type the URL into your browser to access them.

Note: You will need to create a free *NelsonNet* account.

https://get.ga/a-hsc-maths-ext-2

9780170459273

HSC EXAM FORMAT

Mathematics Extension 2 students complete two HSC exams: **Mathematics Extension 1** and **Mathematics Extension 2**.

The following information about the exams was correct at the time of printing in 2021. Please check the NESA website in case it has changed. Visit www.educationstandards.nsw.edu.au, select 'Year 11–Year 12', 'Syllabuses A–Z', 'Mathematics Extension 1/Extension 2', then 'Assessment and Reporting'. Scroll down to 'HSC examination specifications'.

Mathematics Extension 1 HSC exam

	Questions	Marks	Recommended time
Section I	10 multiple-choice questions	10	15 min
Section II	4 multi-part short-answer questions, average 15 marks each, including questions worth 4 or 5 marks	60	1 h 45 min
Total		70	2 h

- Reading time: 10 minutes; use this time to preview the whole exam.

- Working time: 2 hours

- Questions focus on Year 12 outcomes but Year 11 knowledge may be examined.

- Answers are to be written in separate answer booklets.

- A reference sheet is provided at the back of the exam paper and also this book, containing common formulas.

- The 4- or 5-mark questions are usually complex problems that require many steps of working and careful planning.

- Having 2 hours for a total of 70 marks means that you have an average of 1.7 minutes per mark (or approximately 5 minutes for 3 marks).

- If you budget 15 minutes for Section I and 20 minutes per question for Section II, you will then have 25 minutes at the end of the exam to check over your work and/or complete questions you missed.

Mathematics Extension 2 HSC exam

	Questions	Marks	Recommended time
Section I	10 multiple-choice questions	10	15 min
Section II	6 multi-part short-answer questions, average 15 marks each, including questions worth 4 or 5 marks	90	2 h 45 min
Total		100	3 h

- Reading time: 10 minutes

- Working time: 3 hours

- Answers are to be written in separate answer booklets.

- Having 3 hours for a total of 100 marks means that you have an average of 1.8 minutes per mark (or approximately 5 minutes for 3 marks).

- If you budget 15 minutes for Section I and 25 minutes per question for Section II, then you will have 15 minutes at the end of the exam to check over your work and/or complete questions you missed.

STUDY AND EXAM ADVICE

A journey of a thousand miles begins with a single step.
Lao Tzu (c. 570–490 BCE), Chinese philosopher

I've always believed that if you put in the work, the results will come.
Michael Jordan (1963–), American basketball player

Four PRACtical steps for maths study

1. **P**ractise your maths

- Do your homework.
- Learning maths is about mastering a collection of skills.
- You become successful at maths by doing it more, through regular practice and learning.
- Aim to achieve a high level of understanding.

2. **R**ewrite your maths

- Homework and study are not the same thing. Study is your private 'revision' work for strengthening your understanding of a subject.
- Before you begin any questions, make sure you have a thorough understanding of the topic.
- Take ownership of your maths. Rewrite the theory and examples in your own words.
- Summarise each topic to see the 'whole picture' and know it all.

3. **A**ttack your maths

- All maths knowledge is interconnected. If you don't understand one topic fully, then you may have trouble learning another topic.
- Mathematics is not an HSC course you can learn 'by halves' – you have to know it all!
- Fill in any gaps in your mathematical knowledge to see the 'whole picture'.
- Identify your areas of weakness and work on them.
- Spend most of your study time on the topics you find difficult.

4. **C**heck your maths

- After you have mastered a maths skill, such as graphing a quadratic equation, no further learning or reading is needed, just more practice.
- Compared to other subjects, the types of questions asked in maths exams are conventional and predictable.
- Test your understanding with revision exercises, practice papers and past exam papers.
- Develop your exam technique and problem-solving skills.
- Go back to steps 1–3 to improve your study habits.

Topic summaries and concept maps

Summarise each topic when you have completed it, to create useful study notes for revising the course, especially before exams. Use a notebook or folder to list the important ideas, formulas, terminology and skills for each topic. This book is a good study guide, but educational research shows that effective learning takes place when you rewrite learned knowledge in your own words.

A good topic summary runs for 2 to 4 pages. It is a condensed, personalised version of your course notes. This is your interpretation of a topic, so include your own comments, symbols, diagrams, observations and reminders. Highlight important facts using boxes and include a glossary of key words and phrases.

A concept map or mind map is a topic summary in graphic form, with boxes, branches and arrows showing the connections between the main ideas of the topic. This book contains examples of concept maps. The topic name is central to the map, with key concepts or subheadings listing important details and formulas. Concept maps are powerful because they present an overview of a topic on one large sheet of paper. Visual learners absorb and recall information better when they use concept maps.

When compiling a topic summary, use your class notes, your textbook and the companion A+ STUDY NOTES book. Ask your teacher for a copy of the course syllabus or the school's teaching program, which includes the objectives and outcomes of every topic in dot point form.

Attacking your weak areas

Most of your study time should be spent on attacking your weak areas to fill in any gaps in your maths knowledge. Don't spend too much time on work you already know well, unless you need a confidence boost! Ask your teacher, use this book or your textbook to improve the understanding of your weak areas and to practise maths skills. Use your topic summaries for general revision, but spend longer study periods on overcoming any difficulties in your mastery of the course.

Practising with past exam papers

Why is practising with past exam papers such an effective study technique? It allows you to become familiar with the format, style and level of difficulty expected in a HSC exam, as well as the common topic areas tested. Knowing what to expect helps alleviate exam anxiety. Remember, mathematics is a subject in which the exam questions are fairly predictable. The exam writers are not going to ask too many unusual questions. By the time you have worked through many past exam papers, this year's HSC exams won't seem that much different.

Don't throw your old exam papers away. Use them to identify your mistakes and weak areas for further study. Revising topics and then working on mixed questions is a great way to study maths. You might like to complete a past HSC exam paper under timed conditions to improve your exam technique.

Past HSC exam papers are available at the NESA website: visit www.educationstandards.nsw.edu.au and select 'Year 11 – Year 12', 'HSC exam papers'. NESA marking feedback and guidelines can also be viewed there. You can find past HSC exam papers with solutions online, in bookstores, at the Mathematical Association of NSW (www.mansw.nsw.edu.au) and at your school (ask your teacher) or library.

Preparing for an exam

- Make a study plan early; don't leave it until the last minute.
- Read and revise your topic summaries.
- Work on your weak areas and learn from your mistakes.
- Don't spend too much time studying concepts you know already.
- Revise by completing revision exercises and past exam papers or assignments.
- Vary the way you study so that you don't become bored: ask someone to quiz you, voice-record your summary, design a poster or concept map, or explain a concept to someone.
- Anticipate the exam:
  - How many questions will there be?
  - What are the types of questions: multiple-choice, short-answer, long-answer, problem-solving?
  - Which topics will be tested?
  - How many marks are there in each section?
  - How long is the exam?
  - How much time should I spend on each question/section?
  - Which formulas are on the reference sheet and how do I use them in the exam?

During an exam

1. Bring all of your equipment, including a ruler and calculator (check that your calculator works and is in RADIANS mode for trigonometric functions and DEGREES for trigonometric measurements). A highlighter pen may help for tables, graphs and diagrams.

2. Don't worry if you feel nervous before an exam – this is normal and can help you to perform better; however, being too casual or too anxious can harm your performance. Just before the exam begins, take deep, slow breaths to reduce any stress.

3. Write clearly and neatly in black or blue pen, not red. Use a pencil only for diagrams and constructions.

4. Use the **reading time** to browse through the exam to see the work that is ahead of you and the marks allocated to each question. Doing this will ensure you won't miss any questions or pages. Note the harder questions and allow more time for working on them. Leave them if you get stuck, and come back to them later.

5. Attempt every question. It is better to do most of every question and score some marks, rather than ignore questions completely and score 0 for them. Don't leave multiple-choice questions unanswered! Even if you guess, you have a chance of being correct.

6. Easier questions are usually at the beginning, with harder ones at the end. Do an easy question first to boost your confidence. Some students like to leave multiple-choice questions until last so that, if they run out of time, they can make quick guesses. However, some multiple-choice questions can be quite difficult.

7. Read each question and identify what needs to be found and what topic/skill it is testing. The number of marks indicates how much time and working out is required. Highlight any important keywords or clues. Do you need to use the answer to the previous part of the question?

8. After reading each question, and before you start writing, spend a few moments planning and thinking.

9. You don't need to be writing all of the time. What you are writing may be wrong and a waste of time. Spend some time considering the best approach.

10. Make sure each answer seems reasonable and realistic, especially if it involves money or measurement.

11. Show all necessary working, write clearly, draw big diagrams, and set your working out neatly. Write solutions to each part underneath the previous step so that your working out goes down the page, not across.

12. Use a ruler to draw (or read) half-page graphs with labels and axes marked, or to measure scale diagrams.

13. Don't spend too much time on one question. Keep an eye on the time.

14. Make sure you have answered the question. Did you remember to round the answer and/or include units? Did you use all of the relevant information given?

15. If a hard question is taking too long, don't get bogged down. If you're getting nowhere, retrace your steps, start again, or skip the question (circle it) and return to it later with a clearer mind.

16. If you make a mistake, cross it out with a neat line. Don't scribble over it completely or use correction fluid or tape (which is time-consuming and messy). You may still score marks for crossed-out work if it is correct, but don't leave multiple answers! Keep track of your answer booklets and ask for more writing paper if needed.

17. Don't cross out or change an answer too quickly. Research shows that often your first answer is the correct one.

18. Don't round your answer in the middle of a calculation. Round at the end only.

19. Be prepared to write words and sentences in your answers, but don't use abbreviations that you've just made up. Use correct terminology and write one or two sentences for 2 or 3 marks, not mini-essays.

20. If you have time at the end of the exam, double-check your answers, especially for the more difficult questions or questions you are uncertain about.

Ten exam habits of the best HSC students

1. Has clear and careful working and checks their answers

2. Has a strong understanding of basic algebra and calculation

3. Reads (and answers) the whole question

4. Chooses the simplest and quickest method

5. Checks that their answer makes sense or sounds reasonable

6. Draws big, clear diagrams with details and labels

7. Uses a ruler for drawing, measuring and reading graphs

8. Can explain answers in words when needed, in one or two clear sentences

9. Uses the previous parts of a question to solve the next part of the question

10. Rounds answers at the end, not before.

Further resources

Visit the NESA website (www.educationstandards.nsw.edu.au) for the following resources. Select 'Year 11 – Year 12' and then 'Syllabuses A–Z' or 'HSC exam papers'.

- Mathematics Advanced, Extension 1 and Extension 2 syllabuses

- Past HSC exam papers, including marking feedback and guidelines

- Sample HSC questions/exam papers and marking guidelines

Before 2020, 'Mathematics Advanced' was called 'Mathematics' and although 'Mathematics Extension 1' and 'Mathematics Extension 2' have the same name, they were different courses with some topics that no longer exist. For these exam papers, select 'Year 11 – Year 12', 'Resources archive', 'HSC exam papers archive'.

MATHEMATICAL VERBS

A glossary of 'doing words' common in maths problems and HSC exams

analyse
study in detail the parts of a situation

apply
use knowledge or a procedure in a given situation

calculate
See **evaluate**

classify/identify
state the type, name or feature of an item or situation

comment
express an observation or opinion about a result

compare
show how two or more things are similar or different

complete
fill in detail to make a statement, diagram or table correct or finished

construct
draw an accurate diagram

convert
change from one form to another, for example, from a fraction to a decimal, or from kilograms to grams

decrease
make smaller

describe
state the features of a situation

estimate
make an educated guess for a number, measurement or solution, to find roughly or approximately

evaluate/calculate
find the value of a numerical expression, for example, 3×8^2 or $4x + 1$ when $x = 5$

expand
remove brackets in an algebraic expression, for example, expanding $3(2y + 1)$ gives $6y + 3$

explain
describe why or how

give reasons
show the rules or thinking used when solving a problem. *See also* **justify**

graph
display on a number line, number plane or statistical graph

hence find/prove
calculate an answer or demonstrate a result using previous answers or information supplied

identify
See **classify**

increase
make larger

interpret
find meaning in a mathematical result

justify
give reasons or evidence to support your argument or conclusion. *See also* **give reasons**

measure
determine the size of something, for example, using a ruler to determine the length of a pen

prove
See **show/prove that**

recall
remember and state

show/prove that
(in questions where the answer is given) use calculation, procedure or reasoning to demonstrate that an answer or result is true

simplify
express a result such as a ratio or algebraic expression in its most basic, shortest, neatest form

sketch
draw a rough diagram that shows the general shape or ideas (less accurate than **construct**)

solve
find the value(s) of an unknown pronumeral in an equation or inequality

state
See **write**

substitute
replace part of an expression with another, equivalent expression

verify
check that a solution or result is correct, usually by substituting back into an equation or referring back to the problem

write/state
give an answer, formula or result without showing any working or explanation (This usually means that the answer can be found mentally, or in one step)

9780170459273

SYMBOLS AND ABBREVIATIONS

$<$	is less than
$>$	is greater than
$\leq$	is less than or equal to
$\geq$	is greater than or equal to
( )	parentheses, round brackets
[ ]	(square) brackets
{ }	braces
$\pm$	plus or minus
π	pi = 3.14159 …
$\equiv$	is congruent/identical to
$\circ$	degree
$\angle$	angle
Δ	triangle, the discriminant
$\parallel$	is parallel to
$\perp$	is perpendicular to
x^2	x squared, $x \times x$
x^3	x cubed, $x \times x \times x$
$\mathbb{N}$	the set of natural numbers
$\mathbb{Z}$	the set of integers
$\mathbb{Q}$	the set of rational numbers
$\mathbb{R}$	the set of real numbers
$\mathbb{C}$	the set of complex numbers
$\in$	is an element of, belongs to
$:$	such that
$\cup$	union
$\cap$	intersection
∞	infinity
$\lvert x \rvert$	absolute value or magnitude of x
$\overline{z}$	the conjugate of z
$\underset{\sim}{v}$	the vector v
$\overrightarrow{AB}$	the vector AB
$\underset{\sim}{u} \cdot \underset{\sim}{v}$	the scalar product of $\underset{\sim}{u}$ and $\underset{\sim}{v}$
$\text{proj}_{\underset{\sim}{u}} \underset{\sim}{v}$	the projection of $\underset{\sim}{v}$ onto $\underset{\sim}{u}$
$\lim\limits_{h \to 0}$	the limit as $h \to 0$
$\dfrac{dy}{dx}, y', f'(x)$	the first derivative of $y, f(x)$
$\dfrac{d^2y}{dx^2}, y'', f''(x)$	the second derivative of $y, f(x)$
$\dot{x}, v$	$\dfrac{dx}{dt}$, velocity
$\ddot{x}, a$	$\dfrac{d^2x}{dt^2}$, acceleration
$\int f(x)\,dx$	the integral of $f(x)$
$f^{-1}(x)$	the inverse function of $f(x)$
$\sin^{-1}, \arcsin$	the inverse sine function
Σ	sigma, the sum of
$\sum\limits_{r=1}^{n} T_r$	the sum of T_r from $r = 1$ to n
$\therefore$	therefore
$[a, b], a \leq x \leq b$	the interval of x-values from a to b (including a and b)
$(a, b), a < x < b$	the interval of x-values between a and b (excluding a and b)
$P(E)$	the probability of event E occurring
$P(\overline{E})$	the probability of event E not occurring
$A \cup B$	A union B, A or B
$A \cap B$	A intersection B, A and B
$P(A \mid B)$	the probability of A given B
$n!$	n factorial, $n(n-1)(n-2) \ldots \times 1$
$^nC_r, \dbinom{n}{r}$	the number of combinations of r objects from n objects
^nP_r	the number of permutations of r objects from n objects
PDF	probability density function
CDF	cumulative distribution function
$X \sim \text{Bin}(n, p)$	X is a random variable of the binomial distribution
$\hat{p}$	sample proportion
LHS	left-hand side
RHS	right-hand side
p.a.	per annum (per year)
cos	cosine ratio
sin	sine ratio
tan	tangent ratio
$\overline{x}$	the mean
$\mu = E(X)$	the population mean, expected value
σ	the standard deviation
$\text{Var}(X) = \sigma^2$	the variance
Q_1	first quartile or lower quartile
Q_2	median (second quartile)
Q_3	third quartile or upper quartile
IQR	interquartile range
α	alpha
θ	theta
m	gradient
$\forall$	for all
$\exists$	there exists
$P \Rightarrow Q$	if P then Q, P implies Q
iff	if and only if
$P \Leftrightarrow Q$	P if and only if Q
$\neg P$	not P
RTP	required to prove
QED	demonstrated as required

9780170459273

A+ HSC YEAR 12 MATHEMATICS

STUDY NOTES

Authors:

Tania Eastcott
Rachel Eastcott

Sarah Hamper

Karen Man
Ashleigh Della Marta

Jim Green
Janet Hunter

PRACTICE EXAMS

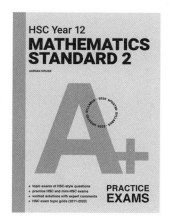   

Authors:

Adrian Kruse

Simon Meli

John Drake

Jim Green
Janet Hunter

Jim Green and **Janet Hunter** also wrote the textbook *Maths in Focus 12 Mathematics Extension 2.*

CHAPTER 1
TOPIC EXAM

Proof

MEX-P1 The nature of proof
MEX-P2 Further proof by mathematical induction

- A reference sheet is provided on page 179 at the back of this book
- For questions in Section II, show relevant mathematical reasoning and/or calculations

Reading time: 4 minutes
Working time: 1 hour
Total marks: 33

Section I – 3 questions, 3 marks
- Attempt Questions 1–3
- Allow about 5 minutes for this section

Section II – 2 questions, 30 marks
- Attempt Questions 4–5
- Allow about 55 minutes for this section

Section I

• Attempt Questions 1–3 • Allow about 5 minutes for this section	**3 marks**

Question 1

What is the converse of the following statement?

'If you are vaccinated, then you do not contract the virus.'

A 'If you contract the virus, then you are not vaccinated.'

B 'If you do not contract the virus, then you are vaccinated.'

C 'If you are not vaccinated, then you do not contract the virus.'

D 'If you are not vaccinated, then you contract the virus.'

Question 2

Consider the statement:

'All students who do Mathematics are logical.'

Which of the following statements is a counterexample to the statement?

A History students are logical.

B All logical students do Mathematics.

C Students who don't do Mathematics are logical.

D Students who are not logical do Mathematics.

Question 3

Which of the following sums is $\displaystyle\sum_{k=1}^{2n} (-1)^k (2k + 1)$?

A $3 - 5 + 7 - \cdots - (4n + 1)$

B $-3 + 5 - 7 + \cdots - (4n + 1)$

C $3 - 5 + 7 - \cdots + (4n + 1)$

D $-3 + 5 - 7 + \cdots + (4n + 1)$

Section II

• Attempt Questions 4–5   **30 marks**
• Allow about 55 minutes for this section
• Answer the questions in the spaces provided. These spaces provide guidance for the expected length of response.
• Your responses should include relevant mathematical reasoning and/or calculations.

Question 4 (15 marks)

a Prove that if M is odd and N is even, where $M > N$, then $M^2 - N^2$ is odd.   3 marks

b Prove by contradiction that $\log_2 3$ is irrational.   3 marks

c Determine whether each statement below is true. If not, give a counterexample.

   i $\forall x, y \in \mathbb{N}, \exists z \in \mathbb{N}: z = \sqrt{x^2 + y^2}$ 2 marks

   ii $\forall a, b \in \mathbb{R}$, if $a > b$, then $a^2 > b^2$.   2 marks

   iii $\forall x \in \mathbb{R}, x^3 - 1$ is divisible by $x - 1$.   2 marks

d Prove each inequality.

   i $\dfrac{x^2 + y^2}{2} \geq xy, \forall x, y \in \mathbb{R}$.   1 mark

ii $\dfrac{a}{b} + \dfrac{b}{a} \geq 2, \forall a, b \in \mathbb{R}, a, b > 0.$

2 marks

Question 5 (15 marks)

a Prove the triangle inequality: $|x + y| \leq |x| + |y|.$

2 marks

b Show that $\dfrac{d^n}{dx^n}\left[\ln(x + 1)\right] = \dfrac{(-1)^{n-1}(n-1)!}{(x+1)^n}, \forall n \in \mathbb{N}.$

3 marks

TOPIC EXAM

c Define the statements P and Q as follows:

P: A quadrilateral has diagonals that bisect each other.
Q: A quadrilateral is a rectangle.

i Write the negation of statement P.   1 mark

ii Consider the statement $P \Rightarrow Q$: 'If a quadrilateral has diagonals that bisect each other, then it is a rectangle.'

Write the contrapositive statement to this statement.   1 mark

iii Is the contrapositive statement in part **ii** true? Give a reason for your answer.   2 marks

iv Hence, explain why $P \Rightarrow Q$ is not true.   1 mark

d Consider the pattern formed by arranging toothpicks:

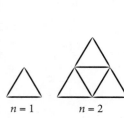 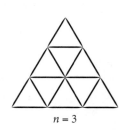 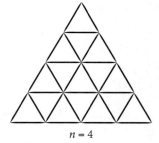

$n = 1$ $n = 2$ $n = 3$ $n = 4$

Side length, n	1	2	3	4	5
Number of toothpicks, T_n	3	9	18	30	

i Determine the number of toothpicks, T_n, required to make a triangle of side length 5.   1 mark

ii The number of toothpicks, T_n, required for a triangle of side length n is given by   1 mark

$$T_n = T_{n-1} + an,$$

where a is a constant.

Find the value of a.

iii Hence, prove by mathematical induction that if $T_1 = 3$ and $T_n = T_{n-1} + 3n$, then: 3 marks

$$T_n = \frac{3n(n+1)}{2} \text{ for all integers } n \geq 1.$$

END OF PAPER

WORKED SOLUTIONS

Section I (1 mark each)

Question 1

B The converse of the statement $P \Rightarrow Q$:

'If you are vaccinated, then you do not contract the virus.'

is $Q \Rightarrow P$: 'If you do not contract the virus, then you are vaccinated.'

> The definition of converse is tested here. Swap the if-then phrases.

Question 2

D 'All students who do Mathematics are logical.'

A counterexample would be a case that disproves the statement, such as one student who does Mathematics is not logical.

'Students who are not logical do Mathematics.'

> This requires careful thought.

Question 3

D Substituting $k = 1, 2, 3, \ldots, 2n$ into the formula yields the correct result.

$$\sum_{k=1}^{2n} (-1)^k (2k + 1) = -3 + 5 - 7 + \cdots + (4n + 1)$$

since $2n$ is even.

Section II (✓ = 1 mark)

Question 4 (15 marks)

a RTP: $M^2 - N^2 = 2X + 1$ for some $X \in \mathbb{N}$.

Proof: Let $M = 2n + 1$, $n \in \mathbb{N}$ and $N = 2k$, $k \in \mathbb{N}$. ✓

Then $M^2 - N^2 = (2n + 1)^2 - (2k)^2$
$$= 4n^2 + 4n + 1 - 4k^2$$
$$= 4(n^2 + n - k^2) + 1 \quad ✓$$
$$= 2X + 1$$

which is odd for some $X \in \mathbb{N}$.   QED. ✓

> This relies on the forms of an odd and an even number.

b RTP by contradiction that $\log_2 3$ is irrational.

Proof: Assume that $\log_2 3$ is rational, that is, assume $\exists p, q \in \mathbb{N}$ such that $\log_2 3 = \dfrac{p}{q}$, where p, q have no common factors, $q \neq 0$. ✓

Then $\log_2 3 = \dfrac{p}{q}$
$$2^{\frac{p}{q}} = 3$$
$$\left(2^{\frac{p}{q}}\right)^q = 3^q$$
$$2^p = 3^q. \quad ✓$$

But since 2 and 3 have no common factors, then no positive integer powers of 2 and 3 can be equal.

OR

Since 2 is even, then 2^p is also even.
Also, since 3 is odd, then 3^q is also odd.

$\therefore 2^p \neq 3^q$.   Contradiction.

So $\log_2 3$ is irrational.   QED. ✓

> Proof of irrationality by contradiction is a common HSC question. Practise this proof.

c **i** $\forall x, y \in \mathbb{N}, \exists z \in \mathbb{N}: z = \sqrt{x^2 + y^2}$

False. ✓

Let $x = 2$ and $y = 3$.

$z = \sqrt{2^2 + 3^2}$

$\quad = \sqrt{13}$

But $\sqrt{13} \notin \mathbb{N}$. ✓

> The quantifier $\forall$ ('for all') means it must be true for every single case. Therefore, you only need one counterexample to disprove the statement.

ii $\forall a, b \in \mathbb{R}$ if $a > b$, then $a^2 > b^2$.

False. ✓

Let $a = 2$ and $b = -3$.

$2 > -3$ but $2^2 \ngtr (-3)^2$. ✓

> This highlights the point that doing the same to both sides of an inequality does not necessarily maintain the truth of the inequality.

iii $\forall x \in \mathbb{R}$, $x^3 - 1$ is divisible by $x - 1$.

True. ✓

By polynomial division,
$x^3 - 1 = (x - 1)(x^2 + x + 1)$ for all x. ✓

> The formula for the difference of 2 cubes $a^3 - b^3$ is a good result to know, even though it is no longer in the Mathematics Advanced course.

d **i** RTP: $\dfrac{x^2 + y^2}{2} \geq xy$, $\forall x, y \in \mathbb{R}$

Proof: Consider the difference
$\dfrac{x^2 + y^2}{2} - xy$.

$\dfrac{x^2 + y^2}{2} - xy = \dfrac{x^2 + y^2 - 2xy}{2}$

$\qquad\qquad\quad = \dfrac{(x - y)^2}{2}$

$\qquad\qquad\quad \geq 0$

Since $(x - y)^2 \geq 0$, $\forall x, y \in \mathbb{R}$.

Therefore, $\dfrac{x^2 + y^2}{2} \geq xy$. QED. ✓

> This is the famous AM-GM (arithmetic mean-geometric mean) inequality. Most inequalities are derived from this basic result.

ii RTP: $\dfrac{a}{b} + \dfrac{b}{a} \geq 2$, $\forall a, b \in \mathbb{R}, a, b > 0$.

Using part **i**, let $x = \sqrt{\dfrac{a}{b}}$ and $y = \sqrt{\dfrac{b}{a}}$

then substituting into $\dfrac{x^2 + y^2}{2} \geq xy$

gives:

$\dfrac{\left(\sqrt{\dfrac{a}{b}}\right)^2 + \left(\sqrt{\dfrac{b}{a}}\right)^2}{2} \geq \sqrt{\dfrac{a}{b}}\sqrt{\dfrac{b}{a}}$ ✓

$\dfrac{\dfrac{a}{b} + \dfrac{b}{a}}{2} \geq \sqrt{\dfrac{a}{b} \times \dfrac{b}{a}}$

$\dfrac{a}{b} + \dfrac{b}{a} \geq 2\sqrt{1}$

$\therefore \dfrac{a}{b} + \dfrac{b}{a} \geq 2$ QED. ✓

> It is possible to use the 'consider the difference' method in part **ii**, but recognising the relationship between the first result for part **i** and the required result here saves time.

Question 5 (15 marks)

a RTP: $|x + y| \leq |x| + |y|$.

Proof: Note that if $\left(|x| + |y|\right)^2 \geq |x + y|^2$
then $|x| + |y| \geq |x + y|$

since both are positive or equal to 0.

Consider

$\left(|x| + |y|\right)^2 = |x|^2 + 2|x||y| + |y|^2$

$\qquad\qquad\quad \geq |x|^2 + 2xy + |y|^2$ ✓

$\qquad\qquad\quad = x^2 + 2xy + y^2$

$\qquad\qquad\quad = (x + y)^2$

$\qquad\qquad\quad = |x + y|^2$

Since both $|x| + |y|$ and $|x + y|$ are positive,
then $|x| + |y| \geq |x + y|$.
OR
$|x + y| \leq |x| + |y|$. QED. ✓

> It is important to know the definition of absolute value.

b RTP: $\dfrac{d^n}{dx^n}[\ln(x+1)] = \dfrac{(-1)^{n-1}(n-1)!}{(x+1)^n}, \forall n \in \mathbb{N}$

Proof:

Consider $\dfrac{d}{dx}[\ln(x+1)] = \dfrac{1}{(x+1)}$

$\qquad\qquad = (x+1)^{-1}.$ ✓

Then $\dfrac{d^2}{dx^2}[\ln(x+1)] = -1(x+1)^{-2} \times 1.$

$\dfrac{d^3}{dx^3}[\ln(x+1)] = (-1) \times (-2)(x+1)^{-3} = (-1)^2 \times 1 \times 2(x+1)^{-3}$ ✓

$\dfrac{d^4}{dx^4}[\ln(x+1)] = (-1) \times (-2) \times (-3)(x+1)^{-4} = (-1)^3 \times 1 \times 2 \times 3(x+1)^{-4}$

$\qquad\qquad\vdots$

$\dfrac{d^n}{dx^n}[\ln(x+1)] = (-1)^{n-1} \times 1 \times 2 \times 3 \times \cdots (n-1)(x+1)^{-n}$

$\qquad\qquad = \dfrac{(-1)^{n-1}(n-1)!}{(x+1)^n}$ ✓

The result appears if we develop the pattern carefully.
Note that the question does not state that this proof could be done by induction.

c i The negation of P is 'not P'.

'A quadrilateral does not have diagonals that bisect each other'. ✓

ii The contrapositive of $P \Rightarrow Q$ is $\neg Q \Rightarrow \neg P$:

'If a quadrilateral is not a rectangle, then it does not have diagonals that bisect each other.' ✓

Negate each part and reverse the order.

iii The statement $\neg P \Rightarrow \neg Q$ is not true. ✓

A parallelogram (or rhombus) has diagonals that bisect each other. ✓

Only one counterexample is required to disprove a statement.

iv A statement and its contrapositive are equivalent, so if the contrapositive is not true, then the original statement is not true. ✓

d i T_n for $n = 5$: we could count the toothpicks or look at the pattern:

Side length, n	1	2	3	4	5
Number of toothpicks, T_n	3	9	18	30	45

$\qquad\qquad$ +6 $\quad$ +9 $\quad$ +12 $\quad$ +15

$T_5 = 30 + 15 = 45.$ ✓

Look at the differences.

ii $T_4 = 30$ and $T_5 = 45$

So $T_n = T_{n-1} + an$

$T_5 = T_4 + a(5)$

$45 = 30 + 5a$

$a = 3.$ ✓

Note the recursive pattern relying on the previous terms.

iii Let $P(n)$ be the proposition that if $T_1 = 3$ and $T_n = T_{n-1} + 3n$, then:

$$T_n = \frac{3n(n+1)}{2} \text{ for all integers } n \geq 1.$$

Proof:

For $P(1)$: $T_1 = 3$ from table and
$T_1 = \frac{3(1)(1+1)}{2} = \frac{6}{2} = 3$ from formula.

$\therefore$ The formula is true. $P(1)$ is true. ✓

Assume $P(k)$ is true:

Given $T_1 = 3$ and $T_k = T_{k-1} + 3k$,

then $T_k = \frac{3k(k+1)}{2}$, for some $k \in \mathbb{N}$ [*]

RTP: $P(k+1)$ is true, that is,

given $T_1 = 3$ and $T_{k+1} = T_k + 3(k+1)$,

then $T_{k+1} = \frac{3(k+1)(k+2)}{2}$. ✓

Proof:

Consider

$$T_{k+1} = T_k + 3(k+1)$$
$$= \frac{3k(k+1)}{2} + 3(k+1) \quad \text{from [*]}$$
$$= \frac{3k(k+1) + 6(k+1)}{2}$$
$$= \frac{3k^2 + 3k + 6k + 6}{2}$$
$$= \frac{3(k^2 + 3k + 2)}{2}$$
$$= \frac{3(k+1)(k+2)}{2}$$
$$= \text{RHS of } P(k+1)$$

So $P(k+1)$ is true.

Therefore, $P(n)$ is true for all integers $n \geq 1$ by mathematical induction. ✓

HSC exam topic grid (2011–2020)

This table shows the coverage of this topic in past HSC exams by question number. The past exams can be downloaded from the NESA website (www.educationstandards.nsw.edu.au) by selecting 'Year 11 – Year 12', 'HSC exam papers'. NESA marking feedback and guidelines can also be found there.

The new Mathematics Extension 2 course was first examined in 2020. For exams before 2020, select 'Year 11 – Year 12', 'Resources archive', 'HSC exam papers archive'.

The language of proof, proof by contradiction and counterexample were introduced to the Mathematics Extension 2 course in 2020.

	The language and methods of proof	Proofs involving numbers and inequalities	Series and divisibility proofs by induction	Inequalities and other proofs by induction
2011	Not in old course	5(b)		3(c)
2012		15(a)	16(b)	
2013		14(a), 16(a)		14(b)
2014		15(a), 16(b)		
2015		15(b), 15(c)		
2016		14(c)		16(c)
2017		13(a)		16(c)
2018		15(c)	16(a)	
2019			14(c)	
2020 new course	7, 8, 14(d), 15(a)	7, 13(c), 14(d)	14(c)	

9780170459273

WORKED SOLUTIONS

CHAPTER 2
TOPIC EXAM

2

3D vectors

MEX-V1 Further work with vectors

    V1.1 Introduction to three-dimensional vectors

    V1.2 Further operations with three-dimensional vectors

    V1.3 Vectors and vector equations of lines

- A reference sheet is provided on page 179 at the back of this book
- For questions in Section II, show relevant mathematical reasoning and/or calculations

Reading time: 4 minutes
Working time: 1 hour
Total marks: 33

Section I – 3 questions, 3 marks
- Attempt Questions 1–3
- Allow about 5 minutes for this section

Section II – 2 questions, 30 marks
- Attempt Questions 4–5
- Allow about 55 minutes for this section

Section I

Question 1

If $\underset{\sim}{u} = \underset{\sim}{i} + 2\underset{\sim}{j} - \underset{\sim}{k}$ and $\underset{\sim}{v} = 2\underset{\sim}{i} + \underset{\sim}{j} - 3\underset{\sim}{k}$, find $2\underset{\sim}{u} - \underset{\sim}{v}$.

A $3\underset{\sim}{i} + 3\underset{\sim}{j} - 4\underset{\sim}{k}$

B $-\underset{\sim}{i} + \underset{\sim}{j} + 2\underset{\sim}{k}$

C $4\underset{\sim}{i} + 5\underset{\sim}{j} - 5\underset{\sim}{k}$

D $3\underset{\sim}{j} + \underset{\sim}{k}$

Question 2

Find the magnitude of $\underset{\sim}{v} = 2\underset{\sim}{i} + 2\underset{\sim}{j} - 3\underset{\sim}{k}$.

A -1

B 1

C $\sqrt{17}$

D 17

Question 3

Which is the vector $\overline{LM}$, joining $L(0, 3, -2)$ to $M(-1, 4, 2)$?

A $3\underset{\sim}{j} - 2\underset{\sim}{k}$

B $-\underset{\sim}{i} + 4\underset{\sim}{j} + 2\underset{\sim}{k}$

C $-\underset{\sim}{i} + \underset{\sim}{j} + 4\underset{\sim}{k}$

D $\underset{\sim}{i} - \underset{\sim}{j} + 4\underset{\sim}{k}$

Section II

- Attempt Questions 4–5   **30 marks**
- Allow about 55 minutes for this section
- Answer the questions in the spaces provided. These spaces provide guidance for the expected length of response.
- Your responses should include relevant mathematical reasoning and/or calculations.

Question 4 (15 marks)

a Using your answer to Question **3**, find the unit vector going from $L(0, 3, -2)$ to $M(-1, 4, 2)$.   2 marks

b Determine the vector $\overrightarrow{MN}$ given that $\overrightarrow{OM} = 2\underset{\sim}{i} - 3\underset{\sim}{j} + \underset{\sim}{k}$ and $\overrightarrow{ON} = \underset{\sim}{i} - 2\underset{\sim}{j} - \underset{\sim}{k}$.   2 marks

c OXY is a triangle where A is the midpoint of OX and B is the midpoint of OY.   3 marks

Show, using vectors, that AB is parallel to XY.

d i Given $u = i + 2j - k$ and $v = 2i + j - 3k$, find $u \cdot v$. 2 marks

ii Find to the nearest degree the angle between the vectors $u = i + 2j - k$ and 2 marks
$v = 2i + j - 3k$.

e Find a vector that is perpendicular to $i - 2j + k$. 2 marks

f Find the value(s) of p if the vectors $i + 2pj + k$ and $-i + 3pj - 2k$ make an angle of 90°. 2 marks

Question 5 (15 marks)

a Determine the radius and centre of the sphere with equation 2 marks
$$x^2 + 2x + y^2 - 6y + z^2 - 4z - 2 = 0.$$

b Describe 2 features of the geometrical relationship between the 2 vectors $\underset{\sim}{u} = \underset{\sim}{i} + 2\underset{\sim}{j} + \underset{\sim}{k}$ 1 mark
and $\underset{\sim}{v} = -2\underset{\sim}{i} - 4\underset{\sim}{j} - 2\underset{\sim}{k}$.

_____

_____

_____

c Find the unit vector of the vector $\underset{\sim}{u} = \underset{\sim}{i} - 2\underset{\sim}{j} - 2\underset{\sim}{k}$.   2 marks

_____

_____

_____

_____

d For the Cartesian equation of a line $\dfrac{x-1}{2} = \dfrac{y+2}{1} = \dfrac{z-2}{-3}$, write the equivalent line   2 marks
in vector form.

_____

_____

_____

_____

_____

e Find the point of intersection of the lines $\begin{pmatrix} 1 \\ 1 \\ 1 \end{pmatrix} + s\begin{pmatrix} 1 \\ 2 \\ -1 \end{pmatrix}$ and $\begin{pmatrix} 3 \\ 2 \\ 1 \end{pmatrix} + t\begin{pmatrix} -1 \\ -5 \\ 3 \end{pmatrix}$.   3 marks

_____

_____

_____

_____

_____

_____

TOPIC EXAM

f **i** A line passes through the points $A(1, 3, -3)$ and $B(2, 1, -5)$.   2 marks

Find the vector equation of the line AB, in the form $\underset{\sim}{r} = \underset{\sim}{a} + \lambda \underset{\sim}{b}$.

ii Determine if the point $C(-1, 7, 1)$ lies on the line AB.   1 mark

g Show that the lines $\begin{pmatrix} 1 \\ 3 \\ -3 \end{pmatrix} + s \begin{pmatrix} 2 \\ 2 \\ -1 \end{pmatrix}$ and $\begin{pmatrix} 4 \\ 7 \\ -3 \end{pmatrix} + t \begin{pmatrix} 1 \\ 2 \\ 2 \end{pmatrix}$ are skew lines.   2 marks

END OF PAPER

WORKED SOLUTIONS

Section I (1 mark each)

Question 1

D $2\underset{\sim}{u} - \underset{\sim}{v} = 2(\underset{\sim}{i} + 2\underset{\sim}{j} - \underset{\sim}{k}) - (2\underset{\sim}{i} + \underset{\sim}{j} - 3\underset{\sim}{k})$

$\qquad\quad = 3\underset{\sim}{j} + \underset{\sim}{k}$

> Be careful when multiplying negatives and collecting terms.

Question 2

C $|v|^2 = 2^2 + 2^2 + (-3)^2 = 17$

$\qquad |v| = \sqrt{17}$

> Don't forget to take the square root.

Question 3

C $\overrightarrow{LM} = (-1 - 0)\underset{\sim}{i} + (4 - 3)\underset{\sim}{j} + (2 - (-2))\underset{\sim}{k}$

$\qquad\quad = -\underset{\sim}{i} + \underset{\sim}{j} + 4\underset{\sim}{k}$

> Make sure that the order of subtraction is correct.

Section II (✓ = 1 mark)

Question 4 (15 marks)

a Let the vector joining the points be $\underset{\sim}{u}$.

$\underset{\sim}{u} = -\underset{\sim}{i} + \underset{\sim}{j} + 4\underset{\sim}{k}$

$|\underset{\sim}{u}| = \sqrt{(-1)^2 + (1)^2 + (4)^2} = \sqrt{18} = 3\sqrt{2}$ ✓

So $\hat{\underset{\sim}{u}} = \dfrac{1}{3\sqrt{2}}(-\underset{\sim}{i} + \underset{\sim}{j} + 4\underset{\sim}{k})$. ✓

> Make sure that you take the square root to find $|\underset{\sim}{u}|$ and use correct notation for the unit vector.

b

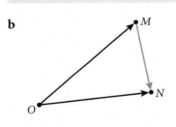

$\overrightarrow{MN} = \overrightarrow{ON} - \overrightarrow{OM}$ ✓

$\overrightarrow{MN} = (\underset{\sim}{i} - 2\underset{\sim}{j} - \underset{\sim}{k}) - (2\underset{\sim}{i} - 3\underset{\sim}{j} + \underset{\sim}{k})$

$\qquad\quad = -\underset{\sim}{i} + \underset{\sim}{j} - 2\underset{\sim}{k}$ ✓

> Check that the order of subtraction is correct and draw a diagram.

c

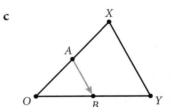

$\overrightarrow{OA} = \dfrac{1}{2}\overrightarrow{OX}$ and $\overrightarrow{OB} = \dfrac{1}{2}\overrightarrow{OY}$

$\overrightarrow{AB} = \overrightarrow{OB} - \overrightarrow{OA}$ ✓

$\qquad = \dfrac{1}{2}\overrightarrow{OY} - \dfrac{1}{2}\overrightarrow{OX}$

$\qquad = \dfrac{1}{2}(\overrightarrow{OY} - \overrightarrow{OX})$

$\qquad = \dfrac{1}{2}\overrightarrow{XY}$ ✓

Hence, $\overrightarrow{AB} \parallel \overrightarrow{XY}$ (and half its length). ✓

> Make sure that your notation and the order of subtraction is correct. Drawing a diagram helps in ensuring your vectors are written and subtracted properly.

d i $\underset{\sim}{u} \cdot \underset{\sim}{v} = x_1 x_2 + y_1 y_2 + z_1 z_2$

$= 1(2) + 2(1) + (-1)(-3)$ ✓

$= 7$ ✓

> Refer to the dot or scalar product – use the notation given and substitute correctly.

ii $\underset{\sim}{u} \cdot \underset{\sim}{v} = |\underset{\sim}{u}||\underset{\sim}{v}|\cos\theta$

$\therefore 7 = \sqrt{6}\sqrt{14}\cos\theta$ ✓

$\cos\theta = \dfrac{7}{\sqrt{84}}$

$\theta \approx 40°$ ✓

> Take the square roots to find $|\underset{\sim}{u}|$ and $|\underset{\sim}{v}|$, then use the correct form of the dot product to find θ.

e For 2 vectors to be perpendicular, $\underset{\sim}{u} \cdot \underset{\sim}{v} = 0$.

$\therefore x_1(1) + y_1(-2) + z_1(1) = 0$ ✓

$\therefore y_1 = \dfrac{x_1 + z_1}{2}$

So, any non-zero vector satisfying this will be perpendicular, for example, $\underset{\sim}{i} + 2\underset{\sim}{j} + 3\underset{\sim}{k}$. ✓

> The expression for y_1 is obtained using the dot product. There are many solutions to this problem as there are infinitely many vectors perpendicular to the given vector.

f $\underset{\sim}{u} \cdot \underset{\sim}{v} = 0$

$1(-1) + 2p(3p) + 1(-2) = 0$ ✓

$6p^2 = 3$

$p^2 = \dfrac{1}{2}$

So $p = \pm\dfrac{1}{\sqrt{2}}$. ✓

> There are 2 possible solutions to $p^2 = \dfrac{1}{2}$ so p can have 2 values.

Question 5 (15 marks)

a Completing the squares:

$(x^2 + 2x + 1) + (y^2 - 6y + 9) + (z^2 - 4z + 4) = 2 + 1 + 9 + 4$ ✓

$(x + 1)^2 + (y - 3)^2 + (z - 2)^2 = 16$

A sphere with centre $(-1, 3, 2)$ and radius 4. ✓

> Use completing the square and don't forget to add to the other side. Remember the general form for a sphere is $(x - a)^2 + (y - b)^2 + (z - c)^2 = r^2$, where the centre is (a, b, c) and the radius is r.

b $\underset{\sim}{v} = -2\underset{\sim}{u}$, or $\underset{\sim}{v}$ is twice as long in the opposite direction. ✓

> You need to write one vector as a scalar multiplied by the other and interpret correctly.

c $\hat{\underset{\sim}{u}} = \dfrac{1}{|\underset{\sim}{u}|}\underset{\sim}{u}$

$|\underset{\sim}{u}| = \sqrt{1^2 + (-2)^2 + (-2)^2} = \sqrt{9} = 3$ ✓

So, $\hat{\underset{\sim}{u}} = \dfrac{1}{3}(\underset{\sim}{i} - 2\underset{\sim}{j} - 2\underset{\sim}{k})$ or $\dfrac{1}{3}\underset{\sim}{i} - \dfrac{2}{3}\underset{\sim}{j} - \dfrac{2}{3}\underset{\sim}{k}$. ✓

> Make sure that you take the square root to find $|\underset{\sim}{u}|$ and use correct notation.

d The direction vector is $2\underset{\sim}{i} + \underset{\sim}{j} - 3\underset{\sim}{k}$ and it passes through the point $(1, -2, 2)$, so the line in vector form is:

$$\underset{\sim}{r} = \begin{pmatrix} 1 \\ -2 \\ 2 \end{pmatrix} + \lambda \begin{pmatrix} 2 \\ 1 \\ -3 \end{pmatrix}. \checkmark$$

> Converting from Cartesian form to vector form of a straight line requires recognition of the direction vector (l, m, n) and a point through which the line passes (a, b, c) from $\dfrac{x - a}{l} = \dfrac{y - b}{m} = \dfrac{z - c}{n}$.

e Using the x-components:

$1 + s = 3 - t$, so $s + t = 2$.

Using the y-components:

$1 + 2s = 2 - 5t$, so $2s + 5t = 1$ $\checkmark$

Solving simultaneously:

$s = 2 - t$

$2(2 - t) + 5t = 1$
$4 - 2t + 5t = 1$
$3t = -3$
$t = -1$

$s = 2 - (-1) = 3$ $\checkmark$

Hence, the point of intersection is $(4, 7, -2)$. $\checkmark$

> You could use x-y, x-z or y-z components to determine s and t using simultaneous equations, no need to test for consistency as these lines do intersect as indicated in the question.

f i The direction vector is $\begin{pmatrix} 2 - 1 \\ 1 - 3 \\ -5 - (-3) \end{pmatrix} = \begin{pmatrix} 1 \\ -2 \\ -2 \end{pmatrix}.$ $\checkmark$

Hence, $\overrightarrow{AB}$ is given by the vector:

$$\underset{\sim}{r} = \begin{pmatrix} 1 \\ 3 \\ -3 \end{pmatrix} + \lambda \begin{pmatrix} 1 \\ -2 \\ -2 \end{pmatrix} \checkmark$$

> Since $\underset{\sim}{r} = \underset{\sim}{a} + \lambda \underset{\sim}{b}$, $\underset{\sim}{a}$ can be either position vector $\overrightarrow{OA}$ or $\overrightarrow{OB}$, but the direction vector is given by $\overrightarrow{OB} - \overrightarrow{OA}$.

ii To test if the point lies on the vector $\overrightarrow{AB}$, we need to find a consistent λ value.

$\therefore 1 + \lambda = -1$, so $\lambda = -2$.

Testing y-components:

$3 - 2\lambda = 7$

Putting $\lambda = -2$:

$3 - 2(-2) = 7$, this is true.

Testing z-components:

$-3 - 2\lambda = 1$

Putting $\lambda = -2$

$-3 - 2(-2) = 1$, this is true. $\checkmark$

> Solving for x-component to establish a value for λ, you must check that this λ works for the y- and z-components too.

g For 2 vectors to not be parallel, one direction vector must not be a multiple of the other.

$$\begin{pmatrix} 2 \\ 2 \\ -1 \end{pmatrix} = k \begin{pmatrix} 1 \\ 2 \\ 2 \end{pmatrix}, \text{ which is not possible } \checkmark$$

To be intersecting, there must be consistent values for s and t.

Equating the x- and y-components:
$1 + 2s = 4 + t$ and $3 + 2s = 7 + 2t$

Solving simultaneously:

$2s - t = 3$
$s - t = 2$

we get $s = 1$, $t = -1$.

These values for s and t work for the y-components but not for the z-components.

$-3 - s = -3 + 2t$ gives $2t + s = 0$ or $s = -2t$, which is not true.

Hence, the lines are skew. $\checkmark$

> To be skew lines, they are neither parallel nor intersecting.

HSC exam topic grid (2020)

This table shows the coverage of this topic in past HSC exams by question number. The past exams can be downloaded from the NESA website (www.educationstandards.nsw.edu.au) by selecting 'Year 11 – Year 12', 'HSC exam papers'. NESA marking feedback and guidelines can also be found there.

The new Mathematics Extension 2 course was first examined in 2020. For exams before 2020, select 'Year 11 – Year 12', 'Resources archive', 'HSC exam papers archive'.

Vectors were introduced to the Mathematics Extension 1 and 2 courses in 2020.

	Operations with vectors	Geometrical proofs	Vector equations of lines	Vector equations of curves and spheres
2020 new course	1, 11(d)	15(b)	3, 11(d), 13(b)	

9780170459273

CHAPTER 3
TOPIC EXAM

Complex numbers

MEX-N1 Introduction to complex numbers

    N1.1 Arithmetic of complex numbers

    N1.2 Geometric representation of a complex number

    N1.3 Other representations of complex numbers

MEX-N2 Using complex numbers

    N2.1 Solving equations with complex numbers

    N2.2 Geometrical implications of complex numbers

- A reference sheet is provided on page 179 at the back of this book
- For questions in Section II, show relevant mathematical reasoning and/or calculations

Reading time: 4 minutes
Working time: 1 hour
Total marks: 33

Section I – 3 questions, 3 marks
- Attempt Questions 1–3
- Allow about 5 minutes for this section

Section II – 2 questions, 30 marks
- Attempt Questions 4–5
- Allow about 55 minutes for this section

9780170459273

Section I

> - Attempt Questions 1–3
> - Allow about 5 minutes for this section
>
> **3 marks**

Question 1

What are the solutions to the equation $z^2 = -\dfrac{1}{2} - \dfrac{i\sqrt{3}}{2}$?

A $\quad z = e^{\pm\frac{\pi}{3}i}$

B $\quad z = \pm e^{\frac{\pi}{3}i}$

C $\quad z = \pm e^{-\frac{\pi}{3}i}$

D $\quad z = \pm e^{\pm\frac{\pi}{3}i}$

Question 2

$P(x) = x^4 + bx^3 + cx^2 + dx + 15$ where $b, c, d \in \mathbb{R}$.

One of the roots of $P(x) = 0$ is $\sqrt{2} + i$.

Which of the following values could be another root?

A $\quad 2 + i$

B $\quad 3 + 2i$

C $\quad 4 + i$

D $\quad 5 + i$

Question 3

Which diagram shows $z = \dfrac{1}{(1 + i)^3}$ on a complex plane?

A
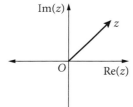

B
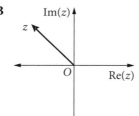

C
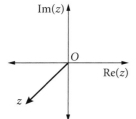

D
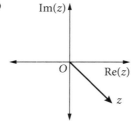

Section II

- Attempt Questions 4–5   **30 marks**
- Allow about 55 minutes for this section
- Answer the questions in the spaces provided. These spaces provide guidance for the expected length of response.
- Your responses should include relevant mathematical reasoning and/or calculations.

Question 4 (15 marks)

a The complex numbers $z = 2(\cos\theta + i\sin\theta)$ and $w = 3(\cos\alpha + i\sin\alpha)$ are plotted below.

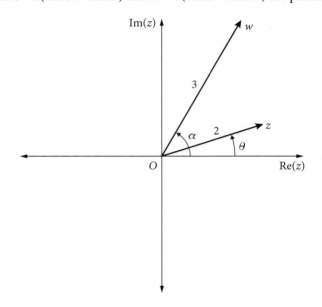

i   Sketch $z + w$ on the diagram.   1 mark

ii Sketch the point V corresponding with the complex number $v = z - w$.   2 marks

iii Sketch $u = iw$.   1 mark

iv Find an expression for $\arg\dfrac{u}{z}$.   1 mark

_____

_____

_____

b Consider the complex numbers $u = \sqrt{2} - i\sqrt{2}$ and $v = 1 + i\sqrt{3}$.

i   Express u and v in modulus-argument form.   2 marks

_____

_____

_____

_____

_____

_____

ii Hence, show that $(uv)^{12}$ is real. 2 marks

_____

_____

_____

_____

_____

c Sketch the graph of each relation on the Argand plane.

i $\left| \dfrac{z + 3}{z - 3i} \right| = 1$ 2 marks

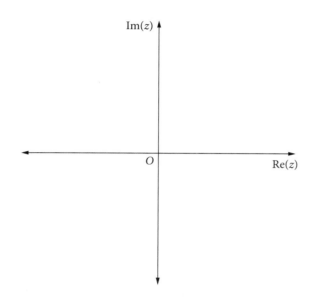

ii $-\dfrac{\pi}{3} \le \text{Arg}(z - 2 + 3i) < \dfrac{3\pi}{4}$ 2 marks

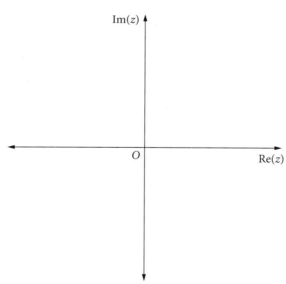

   iii $|\text{Re}(z)| \geq 2$ 2 marks

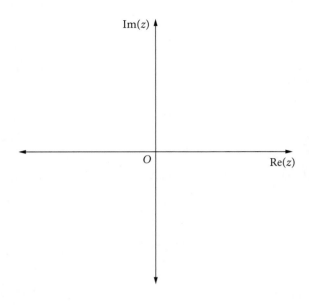

Question 5 (15 marks)

a Find the quadratic equation with roots $2 - i$ and $3 + 2i$, in the form $x^2 + bx + c = 0$,   2 marks
where $b, c \in \mathbb{C}$.

_____

_____

_____

_____

b The complex number β is a root of the equation $z^5 - 1 = 0$.

   i Show the roots of $z^5 - 1 = 0$ on the Argand diagram.   1 mark

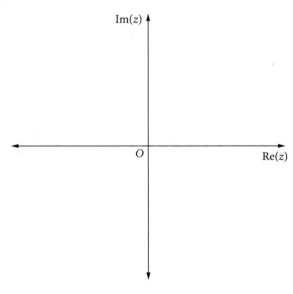

ii Using the sum of the roots of unity, prove that $\cos\dfrac{2\pi}{5} - \cos\dfrac{\pi}{5} = -\dfrac{1}{2}$.   2 marks

iii Factorise $z^5 - 1$ as a product of polynomials with real-valued coefficients.   3 marks

iv Using the result in part **ii**, or otherwise, show that $\cos\dfrac{\pi}{5} = \dfrac{1 + \sqrt{5}}{4}$.   3 marks

TOPIC EXAM

c   ©NESA   2018 HSC EXAM, QUESTION 13(b)

Let $z = 1 - \cos 2\theta + i \sin 2\theta$, where $0 < \theta \le \pi$.

i   Show that $|z| = 2 \sin \theta$.   2 marks

_____

_____

_____

_____

_____

_____

_____

_____

_____

ii   Show that $\arg z = \dfrac{\pi}{2} - \theta$.   2 marks

_____

_____

_____

_____

_____

_____

_____

_____

END OF PAPER

WORKED SOLUTIONS

Section I (1 mark each)

Question 1

C Convert $-\dfrac{1}{2} - \dfrac{i\sqrt{3}}{2}$ to polar/exponential form:

$$r = \sqrt{\left(-\dfrac{1}{2}\right)^2 + \left(-\dfrac{\sqrt{3}}{2}\right)^2}$$

$$= \sqrt{\dfrac{1}{4} + \dfrac{3}{4}}$$

$$= \sqrt{1}$$

$$= 1$$

$$\tan\theta = -\dfrac{\sqrt{3}}{2} \div \left(-\dfrac{1}{2}\right)$$

$$= \sqrt{3}$$

$$\theta = -\dfrac{2\pi}{3} \quad \text{(3rd quadrant)}$$

$$z^2 = -\dfrac{1}{2} - \dfrac{i\sqrt{3}}{2}$$

$$= e^{-\frac{2\pi}{3}i}$$

$$z = \pm\left(e^{-\frac{2\pi}{3}i}\right)^{\frac{1}{2}}$$

$$= \pm e^{-\frac{\pi}{3}i}$$

> Converting the complex number to exponential form allows us to halve the argument, giving us 2 equally spaced roots that are π apart.

Question 2

A As the coefficients are real, then $\sqrt{2} - i$ is also a root. Let the other roots be β and α. The product of the roots is 15:

$$\left(\sqrt{2} - i\right)\left(\sqrt{2} + i\right)\alpha\beta = \dfrac{e}{a}$$

$$(2 + 1)\alpha\beta = \dfrac{15}{1}$$

$$3\alpha\beta = 15$$

$$\alpha\beta = 5$$

The product of the other 2 roots is 5 and as they are complex, they must be conjugate pairs.

$(2 - i)(2 + i) = 5$.

The only option is A.

> This uses the fact that if the coefficients are real, the complex roots come in conjugate pairs.

Question 3

C Convert $1 + i$ to polar form:

$$r = \sqrt{1^2 + 1^2}$$

$$= \sqrt{2}$$

$$\tan\theta = \dfrac{1}{1}$$

$$\theta = \dfrac{\pi}{4}$$

$$z = \dfrac{1}{(1 + i)^3}$$

$$= (1 + i)^{-3}$$

$$= \left[\sqrt{2}\left(\cos\dfrac{\pi}{4} + i\sin\dfrac{\pi}{4}\right)\right]^{-3}$$

$$= \dfrac{1}{2\sqrt{2}}\left[\cos\left(-\dfrac{3\pi}{4}\right) + i\sin\left(-\dfrac{3\pi}{4}\right)\right]$$

$$\arg z = -\dfrac{3\pi}{4}, \text{ (3rd quadrant)}$$

> Express in polar form first, then use the properties of arguments.

Section II ($\checkmark$ = 1 mark)

Question 4 (15 marks)

a

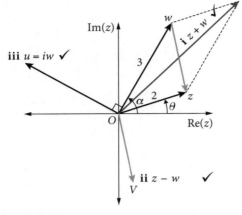

ii V is the point at the head of the vector $z - w$ with tail at the origin $\checkmark$

iii $u = iw$ is w rotated 90° anticlockwise.

iv $\arg \dfrac{u}{z} = \arg u - \arg z$

$$= \alpha + \frac{\pi}{2} - \theta \quad \checkmark$$

This question tests vector addition and multiplication by i. Note the point V is at the head of the vector from O.

---

b i $u = \sqrt{2} - i\sqrt{2}$ and $v = 1 + i\sqrt{3}$

$|u| = \sqrt{\left(\sqrt{2}\right)^2 + \left(-\sqrt{2}\right)^2} = 2$ $|v| = \sqrt{(1)^2 + \left(\sqrt{3}\right)^2} = 2$

$\arg u = -\dfrac{\pi}{4}$ (4th quadrant) $\arg v = \dfrac{\pi}{3}$ (1st quadrant)

$\therefore u = 2\left[\cos\left(-\dfrac{\pi}{4}\right) + i\sin\left(-\dfrac{\pi}{4}\right)\right]$ $\checkmark$ $\therefore v = 2\left(\cos\dfrac{\pi}{3} + i\sin\dfrac{\pi}{3}\right)$ $\checkmark$

Always keep in mind the quadrant to find the correct argument.

---

ii $(uv)^{12} = \left\{2\left[\cos\left(-\dfrac{\pi}{4}\right) + i\sin\left(-\dfrac{\pi}{4}\right)\right] \times 2\left(\cos\dfrac{\pi}{3} + i\sin\dfrac{\pi}{3}\right)\right\}^{12}$

$= \left[2^2\left(\cos\dfrac{\pi}{12} + i\sin\dfrac{\pi}{12}\right)\right]^{12}$ $\checkmark$

$= 2^{24}(\cos\pi + i\sin\pi)$

$= 2^{24}(-1)$

$= -2^{24}$, which is real. $\checkmark$

Multiplication is much easier in polar form.

---

c i $|z + 3| = |z - 3i|$

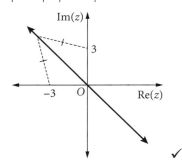

z is equidistant from $3i$ and -3.
The perpendicular bisector is $y = -x$. ✓

z must be equidistant from the two points.

ii $-\dfrac{\pi}{3} \leq \text{Arg}(z - 2 + 3i) < \dfrac{3\pi}{4}$

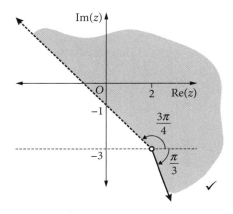

z is the region between the rays from $2 - 3i$.
Open circle at the point. ✓

The arg is measured from the horizontal line pointing right at the point $(2, -3)$. Note the dotted ray and the open circle; the circle is open because arg (0) is undefined.

iii $|\text{Re}(z)| \geq 2$

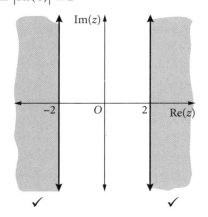

$x \leq -2$ or $x \geq 2$

Recall Re(z) = x.

Question 5 (15 marks)

a The equation will have the form
$x^2 - (\alpha + \beta)x + \alpha\beta = 0$ with roots $2 - i$
and $3 + 2i$. ✓

$\therefore x^2 - (2 - i + 3 + 2i)x + (2 - i)(3 + 2i) = 0$
$x^2 - (5 + i)x + 8 + i = 0$ ✓

$x^2 - (\alpha + \beta)x + \alpha\beta = 0$ is useful to find the quadratic.

b i $z^5 = 1 = \cos(2\pi) + i\sin(2\pi)$

Therefore, one root is

$\beta = \cos\dfrac{2\pi}{5} + i\sin\dfrac{2\pi}{5}.$

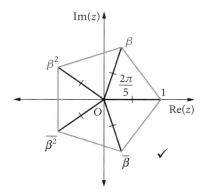

Roots equally spaced from 1.

This is a typical HSC question.

ii RTP: $\cos\dfrac{2\pi}{5} - \cos\dfrac{\pi}{5} = -\dfrac{1}{2}$

Using the sum of roots of unity:

$$\cos\frac{2\pi}{5} + i\sin\frac{2\pi}{5} + \cos\left(-\frac{2\pi}{5}\right) + i\sin\left(-\frac{2\pi}{5}\right) + \cos\frac{4\pi}{5} + i\sin\frac{4\pi}{5} + \cos\left(-\frac{4\pi}{5}\right) + i\sin\left(-\frac{4\pi}{5}\right) + 1 = 0 \quad \checkmark$$

since sum of roots $= -\dfrac{b}{a} = 0$.

$$\therefore \cos\frac{2\pi}{5} + i\sin\frac{2\pi}{5} + \cos\frac{2\pi}{5} - i\sin\frac{2\pi}{5} + \cos\frac{4\pi}{5} + i\sin\frac{4\pi}{5} + \cos\frac{4\pi}{5} - i\sin\frac{4\pi}{5} = -1$$

$$2\cos\frac{2\pi}{5} + 2\cos\frac{4\pi}{5} = -1$$

$$2\cos\frac{2\pi}{5} - 2\cos\frac{\pi}{5} = -1$$

Since $\cos\dfrac{4\pi}{5} = -\cos\dfrac{\pi}{5}$,

$$\cos\frac{2\pi}{5} - \cos\frac{\pi}{5} = -\frac{1}{2} \quad \checkmark$$

The imaginary parts of the conjugates cancel. Note the use of the trigonometric functions properties across quadrants.

iii $z^5 - 1 = (z-1)(z-\beta)(z-\overline{\beta})(z-\beta^2)(z-\overline{\beta^2})$ $\checkmark$

$$= (z-1)[z^2 - (\beta + \overline{\beta})z + \beta\overline{\beta}][z^2 - (\beta^2 + \overline{\beta^2})z + \beta^2\overline{\beta^2}] \quad \checkmark$$

$$= (z-1)\left(z^2 - 2z\cos\frac{2\pi}{5} + 1\right)\left(z^2 - 2z\cos\frac{4\pi}{5} + 1\right) \quad \checkmark$$

Group the conjugates.

iv Now $\cos\dfrac{2\pi}{5} - \cos\dfrac{\pi}{5} = -\dfrac{1}{2}$, then using $\cos 2\theta = 2\cos^2\theta - 1$ and letting $\cos\dfrac{\pi}{5} = x$: $\checkmark$

$$2\cos^2\frac{\pi}{5} - 1 - \cos\frac{\pi}{5} = -\frac{1}{2}$$

$$2x^2 - 1 - x = -\frac{1}{2}$$

$$4x^2 - 2 - 2x = -1$$

$$4x^2 - 2x - 1 = 0$$

$$x = \frac{2 \pm \sqrt{4 - 4(4)(-1)}}{2(4)}$$

$$\cos\frac{\pi}{5} = \frac{2 \pm \sqrt{4 - 4(4)(-1)}}{2(4)}$$

$$= \frac{2 \pm 2\sqrt{5}}{2(4)}$$

$$= \frac{1 \pm \sqrt{5}}{4} \quad \checkmark$$

But $\cos\dfrac{\pi}{5} > 0$ since 1st quadrant, so $\cos\dfrac{\pi}{5} = \dfrac{1+\sqrt{5}}{4}$. $\checkmark$

The connection between the sum of the roots result and the exact value is subtle.

c i $z = 1 - \cos 2\theta + i \sin 2\theta$

$$|z| = \sqrt{(1 - \cos 2\theta)^2 + (\sin 2\theta)^2}$$

$$= \sqrt{1 - 2\cos 2\theta + \cos^2 2\theta + \sin^2 2\theta}$$

$$= \sqrt{1 - 2\cos 2\theta + 1} \quad \checkmark$$

$$= \sqrt{2 - 2\cos 2\theta}$$

$$= \sqrt{2(1 - \cos 2\theta)}$$

$$= \sqrt{2(2\sin^2 \theta)}$$

$$= \sqrt{4\sin^2 \theta}$$

$$= 2\sin \theta \text{ as } \sin \theta \geq 0 \text{ for } 0 < \theta \leq \pi. \quad \checkmark$$

This 2018 HSC exam question tests the definition of modulus and fluency in using the correct trigonometric identities.

ii $\tan(\arg z) = \dfrac{\sin 2\theta}{1 - \cos 2\theta}$

$$= \frac{2 \sin \theta \cos \theta}{2 \sin^2 \theta}$$

$$= \frac{\cos \theta}{\sin \theta} \quad \checkmark$$

$$= \cot \theta$$

$$= \tan\left(\frac{\pi}{2} - \theta\right)$$

As $0 < \theta \leq \pi$, $\dfrac{\pi}{2} - \theta$ is in the range $\left[-\dfrac{\pi}{2}, \dfrac{\pi}{2}\right)$, to prove the result we must check z is in the 1st or 4th quadrant.

Now $\mathrm{Re}(z) = 1 - \cos 2\theta = 2 \sin^2 \theta > 0$, so z is in those quadrants, and:

$$\arg z = \tan^{-1}\left[\tan\left(\frac{\pi}{2} - \theta\right)\right]$$

$$= \frac{\pi}{2} - \theta \quad \checkmark$$

Use of the complementary result, $\cot \theta = \tan\left(\dfrac{\pi}{2} - \theta\right)$, is useful to find the argument. Again, deep knowledge of the trigonometric identities is being tested here, so learn these well and practise using the right ones.

HSC exam topic grid (2011–2020)

This table shows the coverage of this topic in past HSC exams by question number. The past exams can be downloaded from the NESA website (www.educationstandards.nsw.edu.au) by selecting 'Year 11 – Year 12', 'HSC exam papers'. NESA marking feedback and guidelines can also be found there.

The new Mathematics Extension 2 course was first examined in 2020. For exams before 2020, select 'Year 11 – Year 12', 'Resources archive', 'HSC exam papers archive'.

Euler's formula and exponential form were introduced to the Mathematics Extension 2 course in 2020.

	Cartesian and polar form	Euler's formula and exponential form	De Moivre's theorem and equations	Vectors, roots, curves and regions
2011	2(a), (b), (c)	Introduced in 2020	2(d)	4(a), 6(c)
2012	1, 3, 11(a), 11(d)		11(d), 15(b)	11(b), 12(d)
2013	11(a), 14(b), 15(a)		11(c)	3, 5, 11(e)
2014	4, 11(a)		2, 14(a), 15(b)	8, 11(c)
2015	2, 5, 11(a), (b), 12(a)		16(b)(i)	9
2016	10, 11(a), **16(a)**		12(c)	4, 5, **16(a)**, (b)
2017	11(a), **16(a)**		6, 12(b)	1, 3, 11(c), 13(e)
2018	9, 11(a), 11(d), **13(b)**		15(b), 16(c)(iii)	6, 7
2019	1, 8, 11(a), 11(e)		11(e), 16(b)	12(a)
2020 new course	2, 4, 11(a), 14(e)	9, 13(d), 14(a)	4, 11(e)	9

Questions in **bold** can be found in this chapter or in the practice HSC exams in Chapters 8 and 9.

WORKED SOLUTIONS

CHAPTER 4
TOPIC EXAM

4

Further integration

MEX-C1 Further integration

• A reference sheet is provided on page 179 at the back of this book • For questions in Section II, show relevant mathematical reasoning and/or calculations	**Reading time: 4 minutes** **Working time: 1 hour** **Total marks: 33**

Section I – 3 questions, 3 marks
• Attempt Questions 1–3
• Allow about 5 minutes for this section

Section II – 2 questions, 30 marks
• Attempt Questions 4–5
• Allow about 55 minutes for this section

Section I

> - Attempt Questions 1–3   **3 marks**
> - Allow about 5 minutes for this section

Question 1

What substitution would be best to evaluate $\int \cos x \sin^3 x \, dx$?

A $u = \cos x$

B $u = \sin x$

C $u = \cos^3 x$

D $u = \sin^3 x$

Question 2

Find the primitive of $\dfrac{\sec^2(x)}{\tan(x)}$.

A $\dfrac{1}{\tan x} + c$

B $\tan x + c$

C $\ln|\tan x| + c$

D $\sec^2 x + c$

Question 3

Find $\int \dfrac{dx}{x^2 + 4x + 5}$.

A $\tan^{-1}(x + 2) + c$

B $\dfrac{1}{2}\tan^{-1}(x + 2) + c$

C $\tan^{-1}(x + 1) + c$

D $\dfrac{1}{2}\tan^{-1}(x + 1) + c$

9780170459273

Section II

• Attempt Questions 4–5 • Allow about 55 minutes for this section • Answer the questions in the spaces provided. These spaces provide guidance for the expected length of response. • Your responses should include relevant mathematical reasoning and/or calculations.	**30 marks**

Question 4 (15 marks)

a Show that $\int x\sqrt{1-x}\ dx = \frac{2}{5}(1-x)^{\frac{5}{2}} - \frac{2}{3}(1-x)^{\frac{3}{2}} + c$.

3 marks

b Find $\int \frac{2x}{\sqrt{4-x^2}}\ dx$.

3 marks

c i Show that $\frac{1}{x^2-4} = \frac{1}{4(x-2)} - \frac{1}{4(x+2)}$.

3 marks

ii Hence, integrate $\dfrac{1}{x^2 - 4}$ with respect to x.   2 marks

d Evaluate $\displaystyle\int_{4}^{5} \dfrac{2}{x(x-3)}\,dx$.   4 marks

Question 5 (15 marks)

a Show that $\displaystyle\int_{-0.5}^{0.5} \dfrac{dx}{\sqrt{1-2x^2}} = \dfrac{\pi}{2\sqrt{2}}$.   3 marks

b Using integration by parts, show that $\int xe^{2x}\, dx = \frac{1}{4}e^{2x}(2x - 1) + c$. — 3 marks

c Evaluate $\int_0^{\frac{\pi}{2}} x\cos x\, dx$. — 3 marks

d i Let $I_n = \int \sin^n x\, dx$. — 4 marks

Show that $I_n = -\frac{1}{n}\cos x\, \sin^{n-1} x + \frac{n-1}{n}\, I_{n-2}$, for positive integer n.

ii Hence, determine I_3. — 2 marks

END OF PAPER

WORKED SOLUTIONS

Section I (1 mark each)

Question 1

B Using the substitution $u = \sin x$:

$$\int \cos x \sin^3 x \, dx = \int u^3 \, du$$

Question 2

C Putting $u = \tan x$:

$$\int \frac{\sec^2(x)}{\tan(x)} \, dx = \int \frac{1}{u} \, du$$

$$= \ln|u| + c$$

$$= \ln|\tan x| + c$$

'Primitive' is another word for anti-derivative or integral. After solving an integral substitution, write your answer in terms of the original variable (x).

Question 3

A $\displaystyle\int \frac{dx}{x^2 + 4x + 5} = \int \frac{dx}{(x^2 + 4x + 4) + 1}$

$$= \int \frac{dx}{(x + 2)^2 + 1}$$

$$= \tan^{-1}(x + 2) + c$$

Splitting the integrand using partial fractions is not possible, leaving us with completing the square as the only option.

Section II (✓ = 1 mark)

Question 4 (15 marks)

a Putting $u = 1 - x$, $du = -dx$:

$$\int x\sqrt{1 - x} \, dx = \int (1 - u)\sqrt{u} \, (-du) \checkmark$$

$$= \int (u - 1)\sqrt{u} \, du$$

$$= \int (u^{\frac{3}{2}} - u^{\frac{1}{2}}) \, du \checkmark$$

$$= \frac{2}{5}(u)^{\frac{5}{2}} - \frac{2}{3}(u)^{\frac{3}{2}} + c$$

$$= \frac{2}{5}(1 - x)^{\frac{5}{2}} - \frac{2}{3}(1 - x)^{\frac{3}{2}} + c \checkmark$$

Notice that our substitution allows us to assimilate the square root into the rest of the integrand, making our integration easier to do.

b Putting $u = 4 - x^2$, $du = -2x \, dx$: ✓

$$\int \frac{2x}{\sqrt{4 - x^2}} \, dx = \int \frac{-du}{\sqrt{u}}$$

$$= -\int u^{-\frac{1}{2}} \, du \checkmark$$

$$= -2\sqrt{u} + c$$

$$= -2\sqrt{4 - x^2} + c \checkmark$$

The difference between this and a $\sin^{-1}$ integral is the term in the numerator. Make sure that you substitute for dx correctly, change all terms correctly and are careful with integrating $u^{-\frac{1}{2}}$.

c i $\text{RHS} = \dfrac{1}{4(x-2)} - \dfrac{1}{4(x+2)}$

$= \dfrac{(x+2)}{4(x-2)(x+2)} - \dfrac{(x-2)}{4(x+2)(x-2)}$ ✓

$= \dfrac{(x+2)-(x-2)}{4(x-2)(x+2)}$

$= \dfrac{4}{4(x-2)(x+2)}$ ✓

$= \dfrac{1}{x^2-4}$ ✓

$= \text{LHS}$

> With these types of questions, it's much easier to prove RHS = LHS by simplifying than LHS = RHS by partial fractions.

ii $\displaystyle\int \dfrac{1}{x^2-4}\,dx = \int \dfrac{1}{4(x-2)} - \dfrac{1}{4(x+2)}\,dx$

$= \dfrac{1}{4}\displaystyle\int \dfrac{1}{x-2} - \dfrac{1}{x+2}\,dx$ ✓

$= \dfrac{1}{4}\big[\ln|x-2| - \ln|x+2|\big] + c$

$= \dfrac{1}{4}\ln\left|\dfrac{x-2}{x+2}\right| + c$ ✓

> Using the expression in part **i**, factor out the $\frac{1}{4}$ and simplify using logarithm laws. We do not know the limits so use absolute values, as shown.

d Let $\dfrac{2}{x(x-3)} = \dfrac{a}{x} + \dfrac{b}{x-3}$,
which gives $2 = a(x-3) + bx$.

Substituting $x = 3$, $3b = 2$, so $b = \dfrac{2}{3}$. ✓

Substituting $x = 0$, $-3a = 2$, so $a = -\dfrac{2}{3}$.

Hence, we get $\dfrac{2}{x(x-3)} = \dfrac{\frac{2}{3}}{x-3} - \dfrac{\frac{2}{3}}{x}$. ✓

Therefore, $\displaystyle\int_4^5 \dfrac{2}{x(x-3)}\,dx = \int_4^5 \dfrac{\frac{2}{3}}{x-3} - \dfrac{\frac{2}{3}}{x}\,dx$

$= \dfrac{2}{3}\displaystyle\int_4^5 \dfrac{1}{x-3} - \dfrac{1}{x}\,dx$

$= \dfrac{2}{3}\left[\ln\left|\dfrac{x-3}{x}\right|\right]_4^5$ ✓

$= \dfrac{2}{3}\left[\ln\left(\dfrac{2}{5}\right) - \ln\left(\dfrac{1}{4}\right)\right]$

$= \dfrac{2}{3}\ln\left(\dfrac{8}{5}\right).$ ✓

> Using partial fractions to establish decomposition then factoring out the awkward $\frac{2}{3}$ before integration allows for easier evaluation. Since the limits imply that $x - 3$ and x are greater than zero, there is no need for absolute values.

Question 5 (15 marks)

a $\displaystyle\int_{-0.5}^{0.5} \dfrac{dx}{\sqrt{1-2x^2}} = \int_{-\frac{1}{2}}^{\frac{1}{2}} \dfrac{dx}{\sqrt{1-(\sqrt{2}x)^2}}$ ✓

$= \left[\dfrac{1}{\sqrt{2}}\sin^{-1}(\sqrt{2}x)\right]_{-\frac{1}{2}}^{\frac{1}{2}}$ ✓

$= \dfrac{1}{\sqrt{2}}\sin^{-1}\left(\dfrac{1}{\sqrt{2}}\right) - \dfrac{1}{\sqrt{2}}\sin^{-1}\left(-\dfrac{1}{\sqrt{2}}\right)$

$= \dfrac{1}{\sqrt{2}}\left[\dfrac{\pi}{4} - \left(-\dfrac{\pi}{4}\right)\right]$

$= \dfrac{\pi}{2\sqrt{2}}$, as required. ✓

> This is a standard integral
> $$\left(\int \dfrac{dx}{\sqrt{a^2-x^2}} = \sin^{-1}\left(\dfrac{x}{a}\right) + c\right) \text{ after some}$$
> manipulation or substitution, if necessary.

b $\displaystyle\int xe^{2x}\,dx$, putting $u = x$ and $v' = e^{2x}$,
we get $u' = 1$ and $v = \dfrac{1}{2}e^{2x}$. ✓

Using integration by parts:

$\displaystyle\int xe^{2x}\,dx = \dfrac{1}{2}xe^{2x} - \dfrac{1}{2}\int e^{2x}\,dx$ ✓

$= \dfrac{1}{2}xe^{2x} - \dfrac{1}{2}\left(\dfrac{1}{2}e^{2x}\right) + c$

$= \dfrac{1}{4}e^{2x}(2x-1) + c$, as required. ✓

> This is a standard integration by parts. Be careful with multiplication by a negative number and collect terms as required.

c $\displaystyle\int_0^{\frac{\pi}{2}} x\cos x\,dx.$

Putting $u = x$ and $v' = \cos x$,
we get $u' = 1$ and $v = \sin x$. ✓

Using integration by parts:

$\displaystyle\int_0^{\frac{\pi}{2}} x\cos x\,dx = \big[x\sin(x)\big]_0^{\frac{\pi}{2}} - \int_0^{\frac{\pi}{2}}\sin(x)\,dx$ ✓

$= \dfrac{\pi}{2} - \big[-\cos(x)\big]_0^{\frac{\pi}{2}}$

$= \dfrac{\pi}{2} - \big[-0 - (-1)\big]$

$= \dfrac{\pi}{2} - 1$ ✓

> When integrating sin and cos functions, be careful with the negative terms.

9780170459273

d i Let $I_n = \int \sin^n x \, dx$, putting $u = \sin^{n-1} x$ and $v' = \sin x$, ✔

we get $u' = (n-1)\sin^{n-2} x \cos x$ and $v = -\cos x$.

Using integration by parts:

$$\int \sin^n x \, dx = \sin^{n-1} x(-\cos x) - \int (n-1)\sin^{n-2} x \cos x(-\cos x) dx$$

$$= -\cos x \sin^{n-1} x + (n-1)\int \sin^{n-2}(x)\cos^2 x \, dx \ ✔$$

$$= -\cos x \sin^{n-1} x + (n-1)\int \sin^{n-2} x(1 - \sin^2 x) dx$$

$$= -\cos x \sin^{n-1} x + (n-1)\int \sin^{n-2} x - \sin^n x \, dx$$

So,

$$I_n = -\cos x \sin^{n-1} x + (n-1)I_{n-2} - (n-1)I_n \ ✔$$

$$I_n + (n-1)I_n = -\cos x \sin^{n-1} x + (n-1)I_{n-2}$$

$$nI_n = -\cos x \sin^{n-1} x + (n-1)I_{n-2} \ ✔$$

Hence,

$$I_n = -\frac{1}{n}\cos x \sin^{n-1} x + \frac{n-1}{n} I_{n-2}, \text{ as required.}$$

A recurrence relation resulting from integration by parts with trigonometry.

Use the fact that the result is looking for I_{n-2} to assist in choosing a starting point.

ii Hence, $I_3 = -\frac{1}{3}\cos x \sin^2 x + \frac{2}{3}I_1$, where $I_1 = \int \sin x \, dx = -\cos x$. ✔

$I_3 = -\frac{1}{3}\cos x \sin^2 x - \frac{2}{3}\cos x$, or equivalent. ✔

You will need to know I_1, so work that out first.

HSC exam topic grid (2011–2020)

This table shows the coverage of this topic in past HSC exams by question number. The past exams can be downloaded from the NESA website (www.educationstandards.nsw.edu.au) by selecting 'Year 11 – Year 12', 'HSC exam papers'. NESA marking feedback and guidelines can also be found there.

The new Mathematics Extension 2 course was first examined in 2020. For exams before 2020, select 'Year 11 –Year 12', 'Resources archive', 'HSC exam papers archive'.

Euler's formula and exponential form were introduced to the Mathematics Extension 2 course in 2020.

	Integration by substitution	Rational functions and partial fractions	Integration by parts	Integration by recurrence relations
2011	1(b), 1(d), 7(b)	1(c), 1(e)	1(a), 8(a)	8(a)
2012	11(e), 12(a)	11(c), 14(a)	12(c)	12(c)
2013	1, 11(d), 12(a)	6, 11(b)	13(a)	13(a)
2014	7, 10, 13(a), 16(c)	1	11(b)	12(d)
2015	11(f)	11(c)	6, 14(a)	14(a)
2016	14(a)(i), (ii)	12(b), 15(c)(i)	11(b), 12(b)	14(b)
2017	11(d), 11(f)	14(a)	12(c), 15(a)	15(a)
2018	14(a)	1, 11(c), 12(c)	14(c)	14(c)
2019	2, 15(a)	11(c), 11(d), 15(c)	3, 15(c)	15(c)
2020 new course	10	6	11(b), **16(b)**	**16(b)**

Question in **bold** can be found in Practice HSC exam 1, Chapter 8 on pages 102–105.

CHAPTER 5
TOPIC EXAM

Mechanics

MEX-M1 Applications of calculus to mechanics

    M1.1 Simple harmonic motion

    M1.2 Modelling motion without resistance

    M1.3 Resisted motion

    M1.4 Projectiles and resisted motion

- A reference sheet is provided on page 179 at the back of this book
- For questions in Section II, show relevant mathematical reasoning and/or calculations

Reading time: 4 minutes
Working time: 1 hour
Total marks: 33

Section I – 3 questions, 3 marks
- Attempt Questions 1–3
- Allow about 5 minutes for this section

Section II – 2 questions, 30 marks
- Attempt Questions 4–5
- Allow about 55 minutes for this section

9780170459273

Section I

• Attempt Questions 1–3 • Allow about 5 minutes for this section	**3 marks**

Question 1

A particle is moving in simple harmonic motion as described by $x = 2 - 2\cos 2t$.

Over what interval does the particle oscillate?

A $[0, 2]$

B $[0, 4]$

C $[-2, 2]$

D $[-4, 4]$

Question 2

A particle is moving in a straight line. After t seconds, its displacement is x metres from the origin and its velocity is given by $v = 16 + x^2$.

If $x = 4$ initially, which is the correct formula for the time (t) in terms of x?

A $16x + \dfrac{x^3}{3}$

B $16x + \dfrac{x^3}{3} + \dfrac{256}{3}$

C $\tan^{-1}\left(\dfrac{x}{4}\right) - \dfrac{\pi}{4}$

D $\dfrac{1}{4}\tan^{-1}\left(\dfrac{x}{4}\right) - \dfrac{\pi}{16}$

Question 3

A particle of mass 2 kg is projected vertically upwards from $x = 0$.

Its initial velocity is v_0 and the air creates a resistive force of $\dfrac{v^2}{10}$ newtons.

Assuming the acceleration due to gravity is $10\,\text{m}\,\text{s}^{-2}$, what is the maximum height attained by the particle in metres?

A $10\ln\left(\dfrac{200}{200 + v_0^2}\right)$

B $10\ln\left(\dfrac{200 + v_0^2}{200}\right)$

C $10\ln\left(\dfrac{200}{200 - v_0^2}\right)$

D $10\ln\left(\dfrac{200 - v_0^2}{200}\right)$

Section II

• Attempt Questions 4–5 • Allow about 55 minutes for this section • Answer the questions in the spaces provided. These spaces provide guidance for the expected length of response. • Your responses should include relevant mathematical reasoning and/or calculations.	**30 marks**

Question 4 (15 marks)

a Show that a particle whose velocity squared is $v^2 = 96 - 8x - 4x^2$ is undergoing simple harmonic motion and find the centre and period of the motion.

3 marks

b The displacement of a particle from an origin is x metres after t seconds, where

$$x = \sin 2t - \cos 2t.$$

Find the first time ($t > 0$) when the velocity is $2\,\text{m s}^{-1}$.

3 marks

c   A particle moves in a straight line. When its displacement from a fixed origin, O, is   3 marks
x cm, its velocity is v m/s and its acceleration is a m/s^2.

Given that $a = 8x - 2$ and that $v = -1$ when $x = 0$, find $v(x)$.

d   The velocity of a particle is given by $v = \sqrt{2x - 1}$.   3 marks

If the particle is initially at $x = 1$, find an expression for the displacement $x(t)$.

e   A particle moving in a straight line experiences a deceleration of $\dfrac{v^2}{9}$ m/s^2.   3 marks

Initially the particle is projected with a velocity of v_0 m/s.

Show that its velocity is $v = \dfrac{9v_0}{9 + v_0 t}$ m/s.

TOPIC EXAM

Question 5 (15 marks)

a   Show that a particle with position described by $x = 2\cos 2t$ has simple harmonic motion.   2 marks

b   A particle of mass m falls from rest.   2 marks

Assuming that the air resistance per unit mass is k times the speed, prove that the
speed attained by the particle can never exceed $\dfrac{g}{k}$.

c   A projectile is launched from point O at an angle of 25° to a ramp that makes an angle of 8° with the horizontal, as shown in the diagram. The initial velocity of the projectile is 10 m/s.

The projectile hits the ramp at point P. Assume $g \approx 10 \, \text{m/s}^2$.

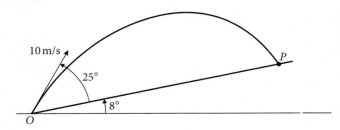

i   Find correct to two decimal places the time it takes for the projectile to hit point P.   4 marks

ii Find correct to one decimal place the distance *OP*.   2 marks

d A skydiver falling through the air at velocity $v\,\mathrm{m\,s^{-1}}$ experiences a resistance of $\frac{1}{6}v$ newtons per unit mass.

    i Find the terminal velocity of the skydiver in terms of *g*.   1 mark

    ii If the effect of a parachute is to increase the resistance to $2v$ newtons per unit mass, calculate in terms of *g* the speed with which the skydiver approaches the ground.   1 mark

TOPIC EXAM

e   A particle of mass 5 kg experiences a resistance, in newtons, of $\frac{1}{6}$ of the square of its   3 marks

velocity in metres per second as it moves through the air. The particle is projected vertically upwards with a velocity of v_0 m/s, and reaches its maximum height at a point P, vertically above the launch position. Assume $g \approx 10 \, \text{ms}^{-2}$.

Find an expression for the time the particle takes to reach the top of its flight.

END OF PAPER

WORKED SOLUTIONS

Section I (1 mark each)

Question 1

B This is the range of $x = 2 - 2\cos 2t$. Because $2\cos 2t$ moves between -2 and 2, $2 - 2\cos 2t$ moves between 0 and 4. It has a minimum of 0 when $t = 0$ and a maximum of 4 when $t = \dfrac{\pi}{2}$.

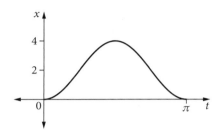

> As the particle is in simple harmonic motion, the end points of the interval through which it oscillates are when $v = 0$. The displacement graph can be drawn to determine these points.

Question 2

D $v = \dfrac{dx}{dt} = 16 + x^2$

$\displaystyle\int \dfrac{dx}{16 + x^2} = \int dt,$

Hence, $\dfrac{1}{4}\tan^{-1}\left(\dfrac{x}{4}\right) = t + c.$

Putting $t = 0$, $x = 4$, we get $c = \dfrac{1}{4}\tan^{-1} 1$

$\qquad\qquad\qquad\qquad = \dfrac{1}{4} \times \dfrac{\pi}{4}$

$\qquad\qquad\qquad\qquad = \dfrac{\pi}{16}.$

So $t = \dfrac{1}{4}\tan^{-1}\left(\dfrac{x}{4}\right) - \dfrac{\pi}{16}.$

> Remember $v = \dfrac{dx}{dt}$, and that we are looking for time (t).

Question 3

B

$2\ddot{x} = -2g - \dfrac{v^2}{10}$

$\ddot{x} = -g - \dfrac{v^2}{20}$

$v\dfrac{dv}{dx} = -10 - \dfrac{v^2}{20}$

$\dfrac{dv}{dx} = -\dfrac{200 + v^2}{20v}$

$10\displaystyle\int \dfrac{2v\,dv}{200 + v^2} = -\int dx$

$10\ln(200 + v^2) = -x + c$

When $t = 0$, $v = v_0$, so $c = 10\ln(200 + v_0{}^2)$

$x = 10\ln(200 + v_0{}^2) - 10\ln(200 + v^2)$

$x = 10\ln\left(\dfrac{200 + v_0{}^2}{200 + v^2}\right)$

Maximum height when $v = 0$, so $10\ln\left(\dfrac{200 + v_0{}^2}{200}\right)$.

> Draw a force diagram to establish the equation of motion.

Section II   ($\checkmark$ = 1 mark)

Question 4 (15 marks)

a $v^2 = 96 - 8x - 4x^2$

$$\frac{1}{2}v^2 = 48 - 4x - 2x^2$$

$$\ddot{x} = \frac{d}{dx}\left(\frac{1}{2}v^2\right) = -4 - 4x = -4(x + 1) \quad \checkmark$$

This is of the form $\ddot{x} = -n^2(x - c)$, which describes simple harmonic motion ($n = 2$), oscillating about $c = -1$. $\checkmark$

Period $T = \dfrac{2\pi}{2} = \pi$ seconds $\checkmark$

> Remember, $\ddot{x} = \dfrac{d\left(\frac{1}{2}v^2\right)}{dx}$ and $\ddot{x} = -n^2(x - c)$, where centre of motion is $x = c$ and the period (T) is given by $T = \dfrac{2\pi}{n}$. The expression for $\ddot{x}$ gives you both the value of c and n required to determine the centre of motion and period.

b $x = \sin 2t - \cos 2t$

Using the auxiliary angle method,

let $\sin 2t - \cos 2t = R\sin(2t - \alpha)$
$$= R(\sin 2t \cos \alpha - \cos 2t \sin \alpha)$$
$$= R\sin 2t \cos \alpha - R\cos 2t \sin \alpha$$

$\therefore R\cos\alpha = 1$ [1]

$\therefore R\sin\alpha = 1$ [2]

[2] ÷ [1]: $\tan\alpha = 1$

$$\alpha = \frac{\pi}{4}$$

$[1]^2 + [2]^2$: $R^2\cos^2\alpha + R^2\sin^2\alpha = 1^2 + 1^2$
$$R^2(\cos^2\alpha + \sin^2\alpha) = 2$$
$$R = \sqrt{2}$$

$\therefore x = \sin 2t - \cos 2t = \sqrt{2}\sin\left(2t - \dfrac{\pi}{4}\right)$

$$\dot{x} = 2\sqrt{2}\cos\left(2t - \frac{\pi}{4}\right) \quad \checkmark$$

When $\dot{x} = 2$, $\cos\left(2t - \dfrac{\pi}{4}\right) = \dfrac{1}{\sqrt{2}}$.

$$2t - \frac{\pi}{4} = -\frac{\pi}{4}, \frac{\pi}{4}, \ldots \quad \checkmark$$

$$2t = 0, \frac{\pi}{2}, \ldots$$

$$t = 0, \frac{\pi}{4}, \ldots$$

Hence, $t = \dfrac{\pi}{4}$, because $t > 0$. $\checkmark$

> It is also possible to differentiate x first as $\dot{x} = 2\cos 2t + 2\sin 2t$, then use the auxiliary angle method.

c $a = \dfrac{d}{dx}\left(\dfrac{1}{2}v^2\right) = 8x - 2$

$$\frac{1}{2}v^2 = 4x^2 - 2x + c \quad \checkmark$$

When $x = 0$, $v = -1$,

$$\therefore c = \frac{1}{2}$$

$$\frac{1}{2}v^2 = 4x^2 - 2x + \frac{1}{2}$$
$$v^2 = 8x^2 - 4x + 1 \quad \checkmark$$
$$v = \pm\sqrt{8x^2 - 4x + 1}, \text{ but when } x = 0, v = -1.$$

So $v = -\sqrt{8x^2 - 4x + 1}$. $\checkmark$

> The aim is to find $v(x)$ so use $\ddot{x} = \dfrac{d\left(\frac{1}{2}v^2\right)}{dx}$ in order to have both the x and v variables in the answer. You need to use the initial conditions to choose either the + or – expression for v.

d $v = \dfrac{dx}{dt} = \sqrt{2x - 1}$

$\dfrac{dt}{dx} = \dfrac{1}{\sqrt{2x - 1}} = (2x - 1)^{-\frac{1}{2}}$ ✓

$t = \dfrac{1}{\frac{1}{2}}(2x - 1)^{\frac{1}{2}} \times \dfrac{1}{2} + c$

$t = \sqrt{2x - 1} + c$

When $t = 0$, $x = 1$. $\therefore c = -1$. ✓

$t = \sqrt{2x - 1} - 1$

$t + 1 = \sqrt{2x - 1}$

$2x - 1 = (t + 1)^2$

$2x = (t + 1)^2 + 1$

So $x = \dfrac{1}{2}\left[(t + 1)^2 + 1\right]$ or equivalent. ✓

> To find $x(t)$, it is necessary to convert v to $\dfrac{dx}{dt}$.
> Be careful when integrating $(2x - 1)^{-\frac{1}{2}}$;
> remember $\int (ax + b)^n \, dx = \dfrac{1}{a}\dfrac{(ax + b)^{n+1}}{n + 1} + c$.

e $m\ddot{x} = -\dfrac{mv^2}{9}$

$\dfrac{dv}{dt} = -\dfrac{v^2}{9}$

$\int \dfrac{dv}{v^2} = -\int \dfrac{dt}{9}$

$-\dfrac{1}{v} = -\dfrac{t}{9} + c$ ✓

When $t = 0$, $v = v_0$, $\therefore c = -\dfrac{1}{v_0}$.

So $\dfrac{1}{v_0} - \dfrac{1}{v} = -\dfrac{t}{9}$ ✓

$\dfrac{1}{v} = \dfrac{1}{v_0} + \dfrac{t}{9}$

$\dfrac{1}{v} = \dfrac{9 + v_0 t}{9 v_0}$ ✓

$v = \dfrac{9 v_0}{9 + v_0 t}$, as required.

> Note that the resistance is acceleration (m/s^2) and so the mass has no effect. You must be careful manipulating to get v_0 as easy mistakes can be made here.

Question 5 (15 marks)

a $x = 2\cos 2t$

$\dot{x} = -4\sin 2t$ ✓

$\ddot{x} = -8\cos 2t$

$= -4 \times 2\cos 2t$

$= -2^2 x$, which is simple harmonic motion (SHM), $\ddot{x} = -n^2 x$, oscillating about the origin. ✓

> SHM is described by $\ddot{x} = -n^2 x$, if oscillating about the origin, or $\ddot{x} = -n^2(x - c)$, if oscillating about $x = c$.

b $m\ddot{x} = mg - mkv$

$\ddot{x} = g - kv$ ✓

When $\ddot{x} = 0$, particle reaches terminal velocity, v_T.

Hence, $g - kv_T = 0$. ✓

$v_T = \dfrac{g}{k}$, therefore, the particle can never exceed $\dfrac{g}{k}$, as required.

> The maximum velocity attained is the terminal velocity, v_T, which occurs when the particle stops accelerating, so $\ddot{x} = 0$.

c **i** $V = 10, \theta = 25° + 8° = 33°$

Deriving the equations of motion:

$\ddot{x} = 0$

$\dot{x} = c$

When $t = 0$, $\dot{x} = V\cos\theta = 10\cos 33°$.

$\dot{x} = 10\cos 33°$

$x = 10t\cos 33° + d$

When $t = 0$, $x = 0$:

   $0 = 0 + d$

$\therefore x = 10t\cos 33°$ [1] ✓

$\ddot{y} = -g = -10$

$\dot{y} = -10t + e$

When $t = 0$, $\dot{y} = V\sin\theta$

$\qquad\qquad\quad = 10\sin 33°$.

$10\sin 33° = 0 + e$

$\dot{y} = -10t + 10\sin 33°$

$y = -5t^2 + 10t\sin 33° + f$

When $t = 0$, $y = 0$:

   $0 = 0 + 0 + f$

$\therefore y = -5t^2 + 10t\sin 33°$ [2] ✓

Let P have coordinates (x, y).

$\tan 8° = \dfrac{y}{x}$ ✓

$x\tan 8° = y$

Substitute [1] and [2] and solve for t:

$$-5t^2 + 10t\sin 33° = 10t\cos 33° \tan 8°$$
$$-5t^2 + 10t(\sin 33° - \cos 33° \tan 8°) = 0$$
$$t - 2(\sin 33° - \cos 33° \tan 8°) = 0 \ (t \neq 0)$$
$$t = 2(\sin 33° - \cos 33° \tan 8°)$$
$$= 0.853\,54\ldots$$
$$\approx 0.85\,\text{s} \ ✓$$

It is necessary to find the coordinates of P and find the equation of the plane on which P lies. There are 2 times when the plane and the trajectory intersect. You can ignore the value of $t = 0$ as this is clearly not where P is located.

ii Using Pythagoras' theorem:

$OP^2 = (10t\cos 33°)^2 + (-5t^2 + 10t\sin 33°)^2$ ✓

Putting $t = 0.853\,54\ldots$

$OP^2 = (8.5354\ldots\cos 33°)^2 + \left[-5(0.85354\ldots)^2 + 8.5354\ldots\sin 33°\right]^2$

$\qquad = 52.25\ldots$

$\quad OP \approx 7.2\,\text{m}$ ✓

Do not forget to find $\sqrt{52.25\ldots}$, a common error.

d i $m\ddot{x} = mg - \dfrac{1}{6}mv$

$\ddot{x} = g - \dfrac{1}{6}v$

Terminal velocity when $\ddot{x} = 0$:

$v_T = 6g$ ✓

A force diagram may help determine the equation of motion. Remember, terminal velocity occurs when acceleration is 0.

ii $m\ddot{x} = mg - 2mv$

$\ddot{x} = g - 2v$

Terminal velocity when $\ddot{x} = 0$:

$v_T = \dfrac{1}{2}g$ ✓

The effect of the parachute is to reduce the speed of approach to the ground so as to land safely. In this case, ≈ 5 m/s is much safer than ≈ 60 m/s without a parachute!

e
$$5\ddot{x} = -5g - \dfrac{1}{6}v^2$$

$$\ddot{x} = -\dfrac{300 + v^2}{30} ✓$$

$$\dfrac{30\,dv}{300 + v^2} = -dt$$

$$\int \dfrac{30\,dv}{300 + v^2} = -\int dt$$

$$30\,\dfrac{1}{\sqrt{300}}\tan^{-1}\left(\dfrac{v}{\sqrt{300}}\right) = -t + c ✓$$

$$30\,\dfrac{1}{10\sqrt{3}}\tan^{-1}\left(\dfrac{v}{10\sqrt{3}}\right) = -t + c$$

$$3\,\dfrac{\sqrt{3}}{3}\tan^{-1}\left(\dfrac{v\sqrt{3}}{30}\right) = -t + c$$

$$\sqrt{3}\tan^{-1}\left(\dfrac{v\sqrt{3}}{30}\right) = -t + c$$

When $t = 0$, $v = v_0$:

$$\sqrt{3}\tan^{-1}\left(\dfrac{v_0\sqrt{3}}{30}\right) = 0 + c$$

$$\therefore t = \sqrt{3}\tan^{-1}\left(\dfrac{v_0\sqrt{3}}{30}\right) - \sqrt{3}\tan^{-1}\left(\dfrac{v\sqrt{3}}{30}\right)$$

The particle reaches the top of flight at $v = 0$.

Time to reach the maximum height is:

$$t = \sqrt{3}\tan^{-1}\left(\dfrac{v_0\sqrt{3}}{30}\right) ✓$$

The resistance is a force in newtons, so no mass needs to be used with the resistance. Be careful rearranging to get an expression to integrate as this is where mistakes may occur.

HSC exam topic grid (2011–2020)

This table shows the coverage of this topic in past HSC exams by Question number. The past exams can be downloaded from the NESA website (www.educationstandards.nsw.edu.au) by selecting 'Year 11 – Year 12', 'HSC exam papers'. NESA marking feedback and guidelines can also be found there.

The new Mathematics Extension 2 course was first examined in 2020. For exams before 2020, select 'Year 11 – Year 12', 'Resources archive', 'HSC exam papers archive'.

Before 2020, simple harmonic motion and velocity/acceleration as functions of x were in the Mathematics Extension 1 course.

	Simple harmonic motion, $v = f(x)$	Modelling motion	Resisted motion	Projectile motion
2011	3(a)*		6(a)	
2012	6*, 13(c)*		13(a)	14(b)*
2013	12(e)*		15(d)	13(c)*
2014	7*, 9, 12(a)*, 12(c)*		14(c)	14(a)*
2015	9*, 13(a)*, 14(b)*		15(a)	14(a)*
2016	13(a)*	15(b)		13(b)*
2017	12(d)*, 13(a)*		13(c)	13(c)*
2018	7*, 10*		14(b)	13(c)*
2019	5*, 12(b)*		14(b)	13(c), 13(d)*
2020 new course	5, 11(c), 13(a)	12(a), 16(a)	14(b)	12(b)

* Mathematics Extension 1 exam

Mathematics Extension 2

PRACTICE MINI-HSC EXAM 1

General instructions	• Reading time: 4 minutes
	• Working time: 1 hour
	• A reference sheet is provided on page 179 at the back of this book
	• For questions in Section II, show relevant mathematical reasoning and/or calculations

Total marks: 33

Section I – 3 questions, 3 marks

• Attempt Questions 1–3

• Allow about 5 minutes for this section

Section II – 2 questions, 30 marks

• Attempt Questions 4–5

• Allow about 55 minutes for this section

Section I

3 marks
Attempt Questions 1–3
Allow about 5 minutes for this section

Circle the correct answer.

Question 1

Consider the square $ABCD$ with the vertices representing the complex numbers α, β, γ, and δ respectively, as shown in the Argand diagram below.

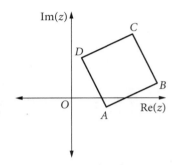

Which equation is correct for γ?

A $\gamma = \delta - 2\alpha + \beta$

B $\gamma = \delta - \alpha + \beta$

C $\gamma = \delta + \beta$

D $\gamma = \delta + \alpha + \beta$

Question 2

Find $\int e^x \sin x \, dx$.

A $e^x \cos x - \int e^x \cos x \, dx$

B $e^x \cos x - \int e^x \sin x \, dx$

C $-e^x \cos x + \int e^x \cos x \, dx$

D $-e^x \cos x + \int e^x \sin x \, dx$

Question 3

Which of the following statements is true?

A $\forall a, b \in \mathbb{R}, \exists c \in \mathbb{Z}: a = bc$

B $\forall a, b \in \mathbb{Z}, \exists c \in \mathbb{N}: a = bc$

C $\forall a, b \in \mathbb{Q}, \exists c \in \mathbb{Z}: a = bc$

D $\forall a, b \in \mathbb{N}, \exists c \in \mathbb{Q}: a = bc$

Section II

30 marks
Attempt Questions 4–5
Allow about 55 minutes for this section

- Answer the questions in the spaces provided. These spaces provide guidance for the expected length of response.
- Your responses should include relevant mathematical reasoning and/or calculations.

Question 4 (15 marks)

a   Find each integral.

   i $\int \dfrac{4x + 1}{x^2 - 2x + 4} \, dx$ 4 marks

   ii $\int_0^{\frac{\sqrt{3}}{2}} \tan^{-1} 2x \, dx$ 3 marks

b   Use De Moivre's theorem to simplify   2 marks

$$\frac{(\cos 2\theta + i \sin 2\theta)^4 (\cos \beta - i \sin \beta)}{(\cos 3\beta + i \sin 3\beta)^{-2}}.$$

Question 4 continues on page 60

Question 4 (continued)

c   Consider the points U and V with corresponding position vectors $\underset{\sim}{u} = \begin{pmatrix} 3 \\ -2 \\ 1 \end{pmatrix}$ and $\underset{\sim}{v} = \begin{pmatrix} -1 \\ 6 \\ 4 \end{pmatrix}$.

i   Find $\text{proj}_v\, \underset{\sim}{u}$, the projection of $\underset{\sim}{u}$ onto $\underset{\sim}{v}$.   2 marks

ii   Hence, find in surd form the distance from the point U to the vector $\underset{\sim}{v}$.   2 marks

iii Find the equation of the line UV in Cartesian form.   2 marks

End of Question 4

Question 5 (15 marks)

a   A particle P is moving in simple harmonic motion. When $t = 0$, P is at the centre of motion, $x = c$, and after 3 seconds, P has moved in the positive direction to $x = 6$. After a further 3 seconds, P is at its maximum displacement, $x = 10$, for the first time.

i   Find the period of the motion.   1 mark

ii   Show that the amplitude is $8 + 4\sqrt{2}$.   3 marks

iii   Find the centre of motion.   1 mark

Question 5 continues on page 62

Question 5 (continued)

b   Let $z = \cos\theta + i\sin\theta$.

    i   Prove that $z^n - z^{-n} = 2i\sin n\theta$.   1 mark

    ii   By considering the binomial expansion of $(z - z^{-1})^5$, show that it is equal to   2 marks

$$2i\sin 5\theta - 10i\sin 3\theta + 20i\sin\theta.$$

    iii   Use the results from parts **i** and **ii** to show that   1 mark

$$\sin^5\theta = \frac{1}{16}\sin 5\theta - \frac{5}{16}\sin 3\theta + \frac{5}{8}\sin\theta.$$

Question 5 continues on page 63

Question 5 (continued)

iv Hence, show that

2 marks

$$\int_{\frac{\pi}{6}}^{\frac{\pi}{3}} \sin^5 \theta \, d\theta = \frac{-203 + 147\sqrt{3}}{480}.$$

Question 5 continues on page 64

Question 5 (continued)

c   Factorise the polynomial $P(z) = z^4 + 4z^3 + 2z^2 + 12z + 45$ over $\mathbb{C}$, given that $P(z) = 0$ has a root   4 marks
of multiplicity 2.

END OF PAPER

WORKED SOLUTIONS

Section I (1 mark each)

Question 1

B $ABCD$ is a square, so the opposite sides are parallel and equal:

$$\gamma - \delta = \beta - \alpha$$

$$\gamma = \beta - \alpha + \delta$$

$$\gamma = \delta - \alpha + \beta$$

There are various ways to approach this question but recognising that $\overrightarrow{DC} = \gamma - \delta$, for instance, is very powerful.

Question 2

C Using integration by parts where $u = e^x$, $v' = \sin x$:

$u' = e^x$, $v = -\cos x$

$$\int e^x \sin x \, dx = -e^x \cos x + \int e^x \cos x \, dx$$

It is possible to swap the functions but none of the other options match this.

Question 3

D If $a = bc$, then $\dfrac{a}{b} = c$ so c might not be a whole number.

$\forall a, b \in \mathbb{N}, \exists c \in \mathbb{Q}: a = bc$ is the only possibility.

Knowing the various sets of numbers is important. Alternatively, you can find counterexamples to options A, B and C.

Section II ($\checkmark$ = 1 mark)

Question 4 (15 marks)

a i $\displaystyle\int \frac{4x + 1}{x^2 - 2x + 4}\, dx = \int \frac{2(2x - 2) + 5}{x^2 - 2x + 4}\, dx$

$\displaystyle \qquad\qquad = \int \frac{2(2x - 2)}{x^2 - 2x + 4} + \frac{5}{(x - 1)^2 + 3}\, dx$ $\checkmark\checkmark$

$\displaystyle \qquad\qquad = 2\ln\left|x^2 - 2x + 4\right| + \frac{5}{\sqrt{3}} \tan^{-1}\left(\frac{x - 1}{\sqrt{3}}\right) + C$

$\qquad\qquad\qquad\qquad \checkmark \qquad\qquad\qquad\quad \checkmark$

> Note the quadratic in the denominator cannot be factorised. This is the clue to splitting the numerator. In this case, because $x^2 - 2x + 4$ is always positive, the absolute-value symbols could be omitted.

ii Method 1: Integrate by parts.

$u = \tan^{-1} 2x \qquad v' = 1$

$u' = \dfrac{2}{1 + 4x^2} \qquad v = x$

$\displaystyle\int_0^{\frac{\sqrt{3}}{2}} \tan^{-1} 2x\, dx = \left[x \tan^{-1} 2x\right]_0^{\frac{\sqrt{3}}{2}} - \int_0^{\frac{\sqrt{3}}{2}} \frac{2x}{1 + 4x^2}\, dx$ $\checkmark$

$\displaystyle \qquad\qquad = \left[x \tan^{-1} 2x - \frac{1}{4}\ln(1 + 4x^2)\right]_0^{\frac{\sqrt{3}}{2}}$

$\displaystyle \qquad\qquad = \left\{\frac{\sqrt{3}}{2} \tan^{-1} \sqrt{3} - \frac{1}{4}\ln\left[1 + 4\left(\frac{3}{4}\right)\right]\right\} - \left[0\tan^{-1} 0 - \frac{1}{4}\ln(1 + 0)\right]$ $\checkmark$

$\displaystyle \qquad\qquad = \left(\frac{\sqrt{3}}{2}\frac{\pi}{3} - \frac{1}{4}\ln 4\right) - (0)$

$\displaystyle \qquad\qquad = \frac{\pi\sqrt{3}}{6} - \ln 4^{\frac{1}{4}}$

$\displaystyle \qquad\qquad = \frac{\pi\sqrt{3}}{6} - \ln(2^2)^{\frac{1}{4}}$

$\displaystyle \qquad\qquad = \frac{\pi\sqrt{3}}{6} - \ln\sqrt{2}$ $\checkmark$

Method 2: Calculate this integral (equivalent to the area under the curve) by subtracting the area between the curve and the y-axis from the rectangle bounded by the limits.

Change the subject and the limits.

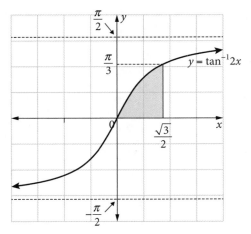

If $y = \tan^{-1} 2x$, then $\frac{1}{2}\tan y = x$.

When $x = \frac{\sqrt{3}}{2}$, $y = \tan^{-1}\left(\frac{\sqrt{3}}{2}\right) = \frac{\pi}{3}$.

$$\int_0^{\frac{\sqrt{3}}{2}} \tan^{-1} 2x \, dx = \text{rectangle area} - \int_0^{\frac{\pi}{3}} \frac{1}{2}\tan y \, dy$$

$$= \left(\frac{\sqrt{3}}{2} \times \frac{\pi}{3}\right) - \frac{1}{2}\int_0^{\frac{\pi}{3}} \frac{\sin y}{\cos y} \, dy \quad \checkmark$$

$$= \frac{\pi\sqrt{3}}{6} + \frac{1}{2}\ln|\cos y|\Big|_0^{\frac{\pi}{3}} \quad \checkmark$$

$$= \frac{\pi\sqrt{3}}{6} + \frac{1}{2}\ln\left|\cos\frac{\pi}{3}\right| - \frac{1}{2}\ln|\cos 0|$$

$$= \frac{\pi\sqrt{3}}{6} + \frac{1}{2}\ln\left(\frac{1}{2}\right) - \frac{1}{2}\ln(1)$$

$$= \frac{\pi\sqrt{3}}{6} - \frac{1}{2}\ln 2 - 0$$

$$= \frac{\pi\sqrt{3}}{6} - \ln 2^{\frac{1}{2}}$$

$$= \frac{\pi\sqrt{3}}{6} - \ln\sqrt{2} \quad \checkmark$$

b $\dfrac{(\cos 2\theta + i\sin 2\theta)^4(\cos\beta - i\sin\beta)}{(\cos 3\beta + i\sin 3\beta)^{-2}} = \dfrac{(\cos 8\theta + i\sin 8\theta)\left[\cos(-\beta) + i\sin(-\beta)\right]}{\left[\cos(-6\beta) + i\sin(-6\beta)\right]}$ $\checkmark$

$$= \cos(8\theta - \beta + 6\beta) + i\sin(8\theta - \beta + 6\beta)$$

$$= \cos(8\theta + 5\beta) + i\sin(8\theta + 5\beta) \quad \checkmark$$

Note the form should be $\cos\alpha + i\sin\alpha$ to use De Moivre's theorem. Use the fact that cos and sin are even and odd functions, respectively, to change the '−' to '+' in the second set of brackets.

c i If $\underset{\sim}{u} = \begin{pmatrix} 3 \\ -2 \\ 1 \end{pmatrix}$ and $\underset{\sim}{v} = \begin{pmatrix} -1 \\ 6 \\ 4 \end{pmatrix}$,

$$\text{proj}_{\underset{\sim}{v}}\, \underset{\sim}{u} = \frac{\underset{\sim}{u}.\underset{\sim}{v}}{|\underset{\sim}{v}|^2}\underset{\sim}{v}$$

$$= \frac{3(-1) - 2(6) + 1(4)}{\left(\sqrt{(-1)^2 + 6^2 + 4^2}\right)^2}\begin{pmatrix} -1 \\ 6 \\ 4 \end{pmatrix} \quad \checkmark$$

$$= -\frac{11}{53}\begin{pmatrix} -1 \\ 6 \\ 4 \end{pmatrix} \quad \checkmark$$

Learn the formula for vector projection thoroughly.

The projection of $\underset{\sim}{u}$ onto $\underset{\sim}{v}$ is given by the formula: $\text{proj}_{\underset{\sim}{v}}\, \underset{\sim}{u} = \dfrac{\underset{\sim}{u}.\underset{\sim}{v}}{|\underset{\sim}{v}|^2}\underset{\sim}{v}$.

ii

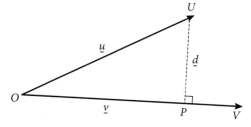

OP is the projection of $\underset{\sim}{u}$ onto $\underset{\sim}{v}$.

Using Pythagoras' theorem,

$$d^2 = \left|\underset{\sim}{u}\right|^2 - \left|\text{proj}_{\underset{\sim}{v}}\underset{\sim}{u}\right|^2 \quad \checkmark$$

$$= \left[3^2 + (-2)^2 + 1^2\right] - \left\{\left(-\frac{11}{53}\right)^2\left[(-1)^2 + 6^2 + 4^2\right]\right\}$$

$$= \frac{621}{53}$$

$$d = \sqrt{\frac{621}{53}} \quad \checkmark$$

> Draw a diagram to see the relationship between the projection and the distance from the point to the line/vector (perpendicular).

iii $\underset{\sim}{v} - \underset{\sim}{u} = \begin{pmatrix} -1 \\ 6 \\ 4 \end{pmatrix} - \begin{pmatrix} 3 \\ -2 \\ 1 \end{pmatrix} = \begin{pmatrix} -4 \\ 8 \\ 3 \end{pmatrix}$

$$\underset{\sim}{r} = \underset{\sim}{u} + \lambda\underset{\sim}{v}$$

$$\begin{pmatrix} x \\ y \\ z \end{pmatrix} = (3, -2, 1) + \lambda\begin{pmatrix} -4 \\ 8 \\ 3 \end{pmatrix} \quad \checkmark$$

$$x = 3 - 4\lambda$$

$$y = -2 + 8\lambda$$

$$z = 1 + 3\lambda$$

$$\therefore \frac{x-3}{-4} = \frac{y+2}{8} = \frac{z-1}{3} \quad \checkmark$$

> To find the equation of the line, find the direction vector between U and V first.

Question 5

a i

A   A

$c - A$ c 6 $c + A$
     $t = 0$ $t = 3$ $t = 6$

Using the fact that a quarter of the time period is taken to go from the centre to an extremity then

Period = $4 \times 6 = 24$ seconds. $\checkmark$

> If we understand the practical nature of SHM, then it simplifies the calculations.

ii Let $x = A\sin(nt) + c$.

Period $= \frac{2\pi}{n} = 24$

$$n = \frac{2\pi}{24} = \frac{\pi}{12}$$

So $x = A\sin\left(\frac{\pi t}{12}\right) + c$. $\checkmark$

We know that when $t = 3$, $x = 6$.

Also, when $t = 6$, $x = 10$.

Substituting:

$$6 = A\sin\left(\frac{3\pi}{12}\right) + c$$

$$6 = A\sin\left(\frac{\pi}{4}\right) + c$$

$$6 = \frac{A}{\sqrt{2}} + c \qquad \qquad \dots[1]$$

$$10 = A\sin\left(\frac{6\pi}{12}\right) + c$$

$$10 = A\sin\left(\frac{\pi}{2}\right) + c$$

$$10 = A + c \qquad \qquad \dots[2] \quad \checkmark$$

Solving simultaneously,

[2] − [1] gives the amplitude:

$$4 = A - \frac{A}{\sqrt{2}}$$

$$= A\left(1 - \frac{1}{\sqrt{2}}\right)$$

$$= A\left(\frac{\sqrt{2}-1}{\sqrt{2}}\right)$$

$$A = \frac{4\sqrt{2}}{\sqrt{2}-1}$$

$$= \frac{4\sqrt{2}}{\sqrt{2}-1} \times \frac{\sqrt{2}+1}{\sqrt{2}+1}$$

$$= \frac{8 + 4\sqrt{2}}{2-1}$$

$$= 8 + 4\sqrt{2} \quad \checkmark$$

> The particle started at the centre, so it is convenient to use a sine function.

iii For the centre, substitute $A = 8 + 4\sqrt{2}$ into [2], or subtract A from 10, the maximum displacement:

$$10 = A + c$$

$$10 = 8 + 4\sqrt{2} + c$$

$$c = 2 - 4\sqrt{2} \quad \checkmark$$

b i $z^n - z^{-n} = (\cos\theta + i\sin\theta)^n - (\cos\theta + i\sin\theta)^{-n}$

$\qquad\qquad = \cos n\theta + i\sin n\theta - [\cos(-n\theta) + i\sin(-n\theta)]$

$\qquad\qquad = \cos n\theta + i\sin n\theta - (\cos n\theta - i\sin n\theta)$

$\qquad\qquad = \cos n\theta + i\sin n\theta - \cos n\theta + i\sin n\theta$

$\qquad\qquad = 2i\sin n\theta$ ✓

This is a well-known identity and a common HSC question, so practise this proof.
Note the use of the trigonometric relations.

ii $(z - z^{-1})^5 = z^5 - 5z^4 z^{-1} + 10z^3 z^{-2} - 10z^2 z^{-3} + 5z^1 z^{-4} - z^{-5}$

$\qquad\qquad = z^5 - 5z^3 + 10z - 10z^{-1} + 5z^{-3} - z^{-5}$

$\qquad\qquad = z^5 - z^{-5} - 5z^3 + 5z^{-3} + 10z - 10z^{-1}$

$\qquad\qquad = (z^5 - z^{-5}) - 5(z^3 - z^{-3}) + 10(z - z^{-1})$ ✓

$\qquad\qquad = (2i\sin 5\theta) - 5(2i\sin 3\theta) + 10(2i\sin\theta)$ (from part **i**)

$\qquad\qquad = 2i\sin 5\theta - 10i\sin 3\theta + 20i\sin\theta$ ✓

Always look for a link to part **i** in HSC exam questions.

iii Applying $n = 5$ to part **i** gives:

$(z - z^{-1})^5 = (2i\sin\theta)^5$

$\qquad\qquad = 32i^5\sin^5\theta$

$\qquad\qquad = 32i\sin^5\theta$ $(i^5 = i^4 \times i = 1 \times i = i)$

Equating our result with part **ii**:

$32i\sin^5\theta = 2i\sin 5\theta - 10i\sin 3\theta + 20i\sin\theta$

$\sin^5\theta = \dfrac{2i\sin 5\theta - 10i\sin 3\theta + 20i\sin\theta}{32i}$

$\sin^5\theta = \dfrac{1}{16}\sin 5\theta - \dfrac{5}{16}\sin 3\theta + \dfrac{5}{8}\sin\theta$ ✓

iv $\displaystyle\int_{\frac{\pi}{6}}^{\frac{\pi}{3}} \sin^5\theta\, d\theta = \int_{\frac{\pi}{6}}^{\frac{\pi}{3}} \dfrac{\sin 5\theta}{16} - \dfrac{5\sin 3\theta}{16} + \dfrac{5\sin\theta}{8}\, d\theta$

$\qquad\qquad = \left[-\dfrac{\cos 5\theta}{80} + \dfrac{5\cos 3\theta}{48} - \dfrac{5\cos\theta}{8} \right]_{\frac{\pi}{6}}^{\frac{\pi}{3}}$ ✓

$\qquad\qquad = \left(-\dfrac{\cos\frac{5\pi}{3}}{80} + \dfrac{5\cos\pi}{48} - \dfrac{5\cos\frac{\pi}{3}}{8} \right) - \left(-\dfrac{\cos\frac{5\pi}{6}}{80} + \dfrac{5\cos\frac{\pi}{2}}{48} - \dfrac{5\cos\frac{\pi}{6}}{8} \right)$

$\qquad\qquad = \left[-\dfrac{\frac{1}{2}}{80} + \dfrac{5(-1)}{48} - \dfrac{5\left(\frac{1}{2}\right)}{8} \right] - \left[-\dfrac{\left(-\frac{\sqrt{3}}{2}\right)}{80} + \dfrac{5(0)}{48} - \dfrac{5\left(\frac{\sqrt{3}}{2}\right)}{8} \right]$

$\qquad\qquad = -\dfrac{1}{160} - \dfrac{5}{48} - \dfrac{5}{16} - \dfrac{\sqrt{3}}{160} + \dfrac{5\sqrt{3}}{16}$

$\qquad\qquad = \dfrac{-203 + 147\sqrt{3}}{480}$ ✓

This is a challenging 6-mark HSC-style question, but notice how all of its parts are related, even though it is not obvious at first. Using the binomial theorem to expand, then to group into $(z - z^{-n})$ pairs, is the key to this question.

c $P(z) = z^4 + 4z^3 + 2z^2 + 12z + 45$ has a double root. We need to find α such that $P(\alpha) = 0$ and $P'(\alpha) = 0$ and α is a factor of 45 as the polynomial is monic.

$$P'(z) = 4z^3 + 12z^2 + 4z + 12 = 0$$
$$4z^2(z + 3) + 4(z + 3) = 0$$
$$(z + 3)(4z^2 + 4) = 0$$
$$4(z + 3)(z^2 + 1) = 0$$
$$z = -3, i, -i \ \checkmark$$

Test $z = -3$ into $P(z)$.

$$P(-3) = (-3)^4 + 4(-3)^3 + 2(-3)^2 + 12(-3) + 45$$
$$= 0$$

So $z = -3$ is a double root and $(z + 3)^2$ is a factor. $\checkmark$

$$\therefore z^4 + 4z^3 + 2x^2 + 12z + 45$$
$$= (z^2 + 6z + 9)(z^2 + Az + B) \text{ where } A, B \in \mathbb{R}.$$

Equating coefficients:

$$9B = 45$$
$$B = 5$$

$$9A + 30 = 12$$
$$9A = -18$$
$$A = -2$$

$$\therefore P(z) = z^4 + 4z^3 + 2z^2 + 12z + 45$$
$$= (z + 3)^2(z^2 - 2z + 5) \ \checkmark$$

Factorising the 2nd factor:

$$z^2 - 2z + 5 = z^2 - 2z + 1 + 4$$
$$= (z - 1)^2 - 4i^2$$
$$= (z - 1 - 2i)(z - 1 + 2i)$$

$$\therefore P(z) = z^4 + 4z^3 + 2z^2 + 12z + 45$$
$$= (z + 3)^2(z - 1 - 2i)(z - 1 + 2i) \ \checkmark$$

> Use the multiple root theorem to identify the first quadratic factor then equate coefficients to find the second. Long division is also possible.
> Note: using the factor theorem alone will not work since there are conjugate complex roots.

Mathematics Extension 2

PRACTICE MINI-HSC EXAM 2

General instructions	• Reading time: 4 minutes
	• Working time: 1 hour
	• A reference sheet is provided on page 179 at the back of this book
	• For questions in Section II, show relevant mathematical reasoning and/or calculations.

Total marks: 33	**Section I – 3 questions, 3 marks**
	• Attempt Questions 1–3
	• Allow about 5 minutes for this section
	Section II – 2 questions, 30 marks
	• Attempt Questions 4–5
	• Allow about 55 minutes for this section

Section I

3 marks
Attempt Questions 1–3
Allow about 5 minutes for this section

Circle the correct answer.

Question 1

In the set of integers, let P be the proposition:

'If $(n + 1)$ is divisible by 3, then $(n^3 + 1)$ is divisible by 3.'

What is the converse statement to P?

A   'If $(n^3 + 1)$ is not divisible by 3, then $(n + 1)$ is not divisible by 3'

B   'If $(n^3 + 1)$ is divisible by 3, then $(n + 1)$ is divisible by 3'

C   'If $(n^3 + 1)$ is not divisible by 3, then $(n + 1)$ is divisible by 3'

D   'If $(n^3 + 1)$ is divisible by 3, then $(n + 1)$ is not divisible by 3'

Question 2

An object of mass m, falling under gravity, experiences a resistance proportional to its velocity v so $R = mkv$.

Which expression describes the terminal velocity of the object?

A $\dfrac{g}{k}$

B $\dfrac{mg}{k}$

C $g - k$

D $g + k$

Question 3

For $\dfrac{2x - 1}{(x - 3)(x + 2)} = \dfrac{A}{x - 3} - \dfrac{B}{x + 2}$, what are the values of A and B?

A $A = 1, B = 1$

B $A = 1, B = -1$

C $A = -1, B = 1$

D $A = -1, B = -1$

Section II

30 marks
Attempt Questions 4–5
Allow about 55 minutes for this section

- Answer the questions in the spaces provided. These spaces provide guidance for the expected length of response.
- Your responses should include relevant mathematical reasoning and/or calculations.

Question 4 (15 marks)

a   Find $\int_0^{\frac{\pi}{2}} x \sin x \, dx$.   3 marks

b   Express $\dfrac{(1-i)^2}{\left(1+i\sqrt{3}\right)^2}$ in the form $re^{i\theta}$.   3 marks

c   Using the substitution $x = \cos^2 \theta$, or otherwise, evaluate $\int_0^{0.5} \sqrt{\dfrac{x}{1-x}} \, dx$.   3 marks

Question 4 continues on page 74

Question 4 (continued)

d   Let *OPQR* be a square on the Argand diagram where *O* is the origin. The points *P* and *R* represent the complex numbers *z* and *iz*, respectively.

    i   Find the complex number represented by *Q*.   1 mark

    ii   The square is rotated through $45°$ in an anticlockwise direction about *O* to $OP'Q'R'$.   2 marks

        Show that P' represents the number $\frac{1}{\sqrt{2}}(1+i)z$.

e   Prove by mathematical induction that $(1+x)^n - 1$ is divisible by x for all integers $n \geq 1$.   3 marks

End of Question 4

Question 5 (15 marks)

a   A particle of mass 1 kg is moving in a horizontal straight line.

It is initially at the origin and travelling with velocity $\sqrt{5}$ m s^{-1}.

The particle is moving with a resisting force $v + v^3$, where v is the velocity.

   i   Show that the acceleration of the particle is given by $\dfrac{dv}{dt} = -(v + v^3)$.   1 mark

   ii   Show that the displacement x of the particle from the origin is given by $x = \tan^{-1}\left(\dfrac{\sqrt{5} - v}{1 + v\sqrt{5}}\right)$.   4 marks

Question 5 continues on page 76

Question 5 (continued)

b   Prove that $\log_2 7$ is irrational.   3 marks

c   Find the point of intersection of the lines $\underset{\sim}{r_1} = \begin{pmatrix} -2 \\ -1 \\ 0 \end{pmatrix} + \lambda \begin{pmatrix} 1 \\ 1 \\ 1 \end{pmatrix}$ and $\underset{\sim}{r_2} = \begin{pmatrix} 8 \\ -6 \\ -11 \end{pmatrix} + \mu \begin{pmatrix} -2 \\ 3 \\ 5 \end{pmatrix}$.   3 marks

Question 5 continues on page 77

Question 5 (continued)

d   Let $I_n = \int_0^1 \dfrac{x^n}{1 + x^2}\, dx$ for $n = 0, 1, 2, \ldots$

    i   Show that $I_n = \dfrac{1}{n-1} - I_{n-2}$, for $n = 2, 3, 4, \ldots$ 3 marks

    ii   Hence, find the value of I_3.   1 mark

END OF PAPER

WORKED SOLUTIONS

Section I (1 mark each)

Question 1

B Converse statement: If $P \Rightarrow Q$, then $Q \Rightarrow P$.

So, in the proposition given:

'If $(n + 1)$ is divisible by 3, then $(n^3 + 1)$ is divisible by 3', the converse statement is:

'If $(n^3 + 1)$ is divisible by 3, then $(n + 1)$ is divisible by 3'.

> Know the definitions of the converse, contrapositive and negation of a given proposition.

Question 2

A $m\ddot{x} = mg - mkv$

$\ddot{x} = g - kv$

Terminal velocity when $\ddot{x} = 0$,

$\therefore kv = g$

So $v = \dfrac{g}{k}$.

> Know how to find the terminal velocity in a variety of circumstances, usually by setting the acceleration to 0.

Question 3

B $\dfrac{2x - 1}{(x - 3)(x + 2)} = \dfrac{A}{x - 3} - \dfrac{B}{x + 2}$

$\therefore 2x - 1 = A(x + 2) - B(x - 3)$

Putting $x = -2$:

$-5 = 5B$

$B = -1$

Putting $x = 3$:

$5 = 5A$

$A = 1$

$\therefore A = 1, B = -1$

> With partial fractions, A and B can be found by substituting values (like above) or expanding and equating coefficients. Learn to use both methods.

Section II ($\checkmark$ = 1 mark)

Question 4 (15 marks)

a Using integration by parts:

Let $u = x$ and $v' = \sin x$

$\therefore u' = 1, v = -\cos x.$ $\checkmark$

$$\int_0^{\frac{\pi}{2}} x \sin x \, dx = \left[-x \cos x\right]_0^{\frac{\pi}{2}} + \int_0^{\frac{\pi}{2}} \cos x \, dx \checkmark$$

$$= (0 - -0) + \left[\sin x\right]_0^{\frac{\pi}{2}}$$

$$= 1 - 0$$

$$= 1 \checkmark$$

> Straightforward integration by parts. If in doubt about substituting limits, use a calculator.

b Convert $1 - i$ to exponential form:

$r = \sqrt{1^2 + (-1)^2}$

$\quad = \sqrt{2}$

$\tan \theta = \dfrac{-1}{1}$

$\quad \theta = -\dfrac{\pi}{4}$

Convert $1 + i\sqrt{3}$ to exponential form:

$r = \sqrt{1^2 + \left(\sqrt{3}\right)^2}$

$\quad = \sqrt{4}$

$\quad = 2$

$\tan \theta = \dfrac{\sqrt{3}}{1} = \sqrt{3}$

$\quad \theta = \dfrac{\pi}{3}$

$\therefore (1 - i) = \sqrt{2}e^{-\frac{i\pi}{4}}$ and $\left(1 + i\sqrt{3}\right) = 2e^{\frac{i\pi}{3}}$ $\checkmark$

$\dfrac{(1-i)^2}{\left(1 + i\sqrt{3}\right)^2} = \dfrac{\left(\sqrt{2}e^{-\frac{i\pi}{4}}\right)^2}{\left(2e^{\frac{i\pi}{3}}\right)^2}$

$\quad = \dfrac{1}{2}e^{-i\left(\frac{\pi}{2} + \frac{2\pi}{3}\right)}$ $\checkmark$

$\quad = \dfrac{1}{2}e^{-i\left(\frac{7\pi}{6}\right)}$ $\left[\text{or } \dfrac{1}{2}e^{i\left(\frac{5\pi}{6}\right)}\right]$ $\checkmark$

> Convert the numerator and denominator into the form $re^{i\theta}$. This allows you to use index laws to simplify.

c Using the substitution $x = \cos^2 \theta$,

$dx = 2 \cos \theta (-\sin \theta) \, d\theta = -2 \sin \theta \cos \theta \, d\theta$

Limit change:

$x = 0, \theta = \dfrac{\pi}{2}$ and $x = 0.5, \theta = \dfrac{\pi}{4}$ $\checkmark$

$$\int_0^{0.5} \sqrt{\frac{x}{1-x}} \, dx = \int_{\frac{\pi}{2}}^{\frac{\pi}{4}} \sqrt{\frac{\cos^2 \theta}{1 - \cos^2 \theta}} (-2 \sin \theta \cos \theta) \, d\theta$$

$$= \int_{\frac{\pi}{2}}^{\frac{\pi}{4}} \frac{\cos \theta}{\sin \theta} (-2 \sin \theta \cos \theta) \, d\theta$$

$$= \int_{\frac{\pi}{2}}^{\frac{\pi}{4}} -2 \cos^2 \theta \, d\theta$$

$$= 2 \int_{\frac{\pi}{4}}^{\frac{\pi}{2}} \cos^2 \theta \, d\theta$$

$$= 2 \int_{\frac{\pi}{4}}^{\frac{\pi}{2}} \frac{1}{2}(1 + \cos 2\theta) \, d\theta \checkmark$$

$$= \left[\theta + \frac{1}{2} \sin 2\theta\right]_{\frac{\pi}{4}}^{\frac{\pi}{2}}$$

$$= \left(\frac{\pi}{2} + 0\right) - \left(\frac{\pi}{4} + \frac{1}{2}[1]\right)$$

$$= \frac{\pi}{4} - \frac{1}{2} \checkmark$$

> This is quite a complex substitution and simplification. As well as the limits needing to be changed, the dx into $d\theta$ needs to be done carefully. Finally, evaluating the definite integral is best left in exact form.

d i $\overrightarrow{OQ} = \overrightarrow{OP} + \overrightarrow{OR}$

$\quad = z + iz$

$\quad = (1 + i)z$ $\checkmark$

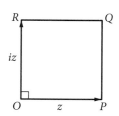

> The diagonal of a rectangle (square) is the sum of the adjacent vectors from O:
> $$\overrightarrow{OQ} = \overrightarrow{OP} + \overrightarrow{PQ} = \overrightarrow{OP} + \overrightarrow{OR}$$

ii OP' is OP rotated $\dfrac{\pi}{4}$ in an anticlockwise direction.

So P' represents the number z multiplied by $1\left(\cos\dfrac{\pi}{4} + i\sin\dfrac{\pi}{4}\right)$ ✓ $=\left(\dfrac{1}{\sqrt{2}} + i\dfrac{1}{\sqrt{2}}\right)$.

So P' represents $\dfrac{z}{\sqrt{2}}(1 + i) = \dfrac{1}{\sqrt{2}}(1 + i)z$. ✓

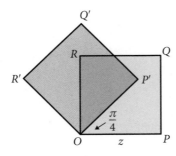

Working in polar or exponential form makes the rotation simpler.

e For $n = 1$:

LHS $= (1 + x)^1 - 1$

      $= x$, which is divisible by x.

Hence, statement is true for $n = 1$. ✓

Assume statement is true for $n = k$, that is, $(1 + x)^k - 1 = xQ$, where Q is a polynomial.   [*] ✓

Then for $n = k + 1$, you need to show that $(1 + x)^{k+1} - 1$ is divisible by x.

LHS $= (1 + x)^{k+1} - 1$

      $= (1 + x)^k(1 + x)^1 - 1$

      $= (xQ + 1)(1 + x) - 1$ by [*]

      $= xQ + x^2Q + 1 + x - 1$

      $= xQ + x^2Q + x$

      $= x[Q + xQ + 1]$, which is divisible by x ✓

Hence, if the statement is true for $n = k$, then it is true for $n = k + 1$.

Since it is true for $n = 1$, then by mathematical induction, it is true for all integers $n \geq 1$.

It is always a good idea to write down the result you are seeking when $n = k + 1$.

Question 5 (15 marks)

a **i** $m\ddot{x} = -(v + v^3)$

Since $m = 1$, we get:

$\ddot{x} = \dfrac{dv}{dt} = -(v + v^3)$, as required. ✓

The particle has unit mass and is slowing down. This is resisted motion in a horizontal line.

ii $v\dfrac{dv}{dx} = -(v + v^3)$ ✓

$\dfrac{v\,dv}{v + v^3} = -dx$

Integrating:

$\displaystyle\int \dfrac{v\,dv}{v + v^3} = -\int dx$

$\displaystyle\int \dfrac{dv}{1 + v^2} = -\int dx$

$\tan^{-1}v = -x + c$ ✓

When $t = 0$, $x = 0$ and $v = \sqrt{5}$:

$\tan^{-1}\sqrt{5} = -0 + c$

$c = \tan^{-1}\sqrt{5}$ ✓

$\therefore x = \tan^{-1}\sqrt{5} - \tan^{-1}v$

$\tan x = \tan[\tan^{-1}\sqrt{5} - \tan^{-1}v]$

$\qquad = \dfrac{\tan\left(\tan^{-1}\sqrt{5}\right) - \tan\left(\tan^{-1}v\right)}{1 + \tan\left(\tan^{-1}\sqrt{5}\right)\tan\left(\tan^{-1}v\right)}$ using $\tan(A - B) = \dfrac{\tan A - \tan B}{1 + \tan A \tan B}$

$\qquad = \dfrac{\sqrt{5} - v}{1 + v\sqrt{5}}$ with $A = \tan^{-1}\sqrt{5}$ and $B = \tan^{-1}v$

$x = \tan^{-1}\left(\dfrac{\sqrt{5} - v}{1 + v\sqrt{5}}\right)$, as required. ✓

> This HSC-style exam question combines integration to find equations of motion with trigonometric identities.

b   Proof by contradiction.

Assume that $\log_2 7$ is rational, that is, assume $\exists\, p, q \in \mathbb{N}$ such that $\log_2 7 = \dfrac{p}{q}$, where p, q have no common factors, $q \neq 0$. ✓

Then $\log_2 7 = \dfrac{p}{q}$:

$2^{\frac{p}{q}} = 7$

$\left(2^{\frac{p}{q}}\right)^q = 7^q$

$2^p = 7^q$ ✓

Since 2 is even, then 2^p is also even.

Since 7 is odd, then 7^q is also odd.

OR

Since 2 and 7 have no common factors, then no powers of 2 and 7 can be equal.

$\therefore 2^p \neq 7^q$. Contradiction.

So $\log_2 7$ is irrational.   QED. ✓

> You need to set up the method of proof by contradiction, that is, assume that $\log_2 7$ is rational and prove otherwise. This is a commonly-asked question on HSC exams.

c Equating components:

$-2 + \lambda = 8 - 2\mu$ $\therefore 2\mu + \lambda = 10$ [1]
$-1 + \lambda = -6 + 3\mu$ $\therefore 3\mu - \lambda = 5$ [2] ✓

Solving simultaneously,

[1] + [2]:

$5\mu = 15$
$\mu = 3$

Substitute into [1]:

$2(3) + \lambda = 10$
$6 + \lambda = 10$
$\lambda = 4$ ✓

Checking for consistency:

$\lambda = -11 + 5\mu$

Putting $\mu = 3$, $\lambda = 4$, as required.

Therefore, point of intersection is $(2, 3, 4)$. ✓

> When finding the point of intersection, the consistency check ensures that the lines do in fact intersect and it does not matter whether you substitute λ or μ to establish the point of intersection.

d i $I_n = \int_0^1 x^{n-2} \dfrac{x^2}{1 + x^2}\, dx$ ✓

$= \int_0^1 x^{n-2} \dfrac{1 + x^2 - 1}{1 + x^2}\, dx$

$= \int_0^1 x^{n-2} \left(1 - \dfrac{1}{1 + x^2}\right) dx$ ✓

$= \int_0^1 x^{n-2}\, dx - \int_0^1 \dfrac{x^{n-2}}{1 + x^2}\, dx$

$= \left[\dfrac{x^{n-1}}{n-1}\right]_0^1 - \int_0^1 \dfrac{x^{n-2}}{1 + x^2}\, dx$ ✓

$= \left[\dfrac{1 - 0}{n - 1}\right] - I_{n-2}$

$= \dfrac{1}{n - 1} - I_{n-2}$, as required.

> Using the fact that the required result has I_{n-2} hints that $\dfrac{x^{n-2}}{1 + x^2}$ must be in the expansion somewhere. This type of problem is harder as you need to manipulate the given expression to another form that is more useful in this circumstance.

ii $I_3 = \dfrac{1}{3 - 1} - I_1$

$= \dfrac{1}{2} - I_1$

$I_1 = \int_0^1 \dfrac{x}{1 + x^2}\, dx$

$= \left[\dfrac{1}{2}\ln(1 + x^2)\right]_0^1$

$= \dfrac{1}{2}\ln 2$

$\therefore I_3 = \dfrac{1}{2} - \dfrac{1}{2}\ln 2$

$= \dfrac{1}{2}(1 - \ln 2)$ ✓

> On simplifying, I_1 needs to be found. This is generally not too difficult.

Mathematics Extension 2

PRACTICE HSC EXAM 1

General instructions	• Reading time: 10 minutes
	• Working time: 3 hours
	• A reference sheet is provided on page 179 at the back of this book
	• For questions in Section II, show relevant mathematical reasoning and/or calculations
Total marks: 100	**Section I – 10 questions, 10 marks**
	• Attempt Questions 1–10
	• Allow about 15 minutes for this section
	Section II – 6 questions, 90 marks
	• Attempt Questions 11–16
	• Allow about 2 hours 45 minutes for this section

Section I

10 marks
Attempt Questions 1–10
Allow about 15 minutes for this section

Circle the correct answer.

Question 1

What are the factors of $z^4 - 16$ over the set of complex numbers?

A $z - 2i, z + 2i$

B $z - 2, z + 2, z - 2i, z + 2i$

C $z^2 - 2, z^2 + 8$

D $2, -2, 2i, -2i$

Question 2

A particle travelling along a straight line has a velocity given by $v = 2\tan^{-1}\left(\frac{t}{4}\right)$, where v is the velocity in $\mathrm{m\,s}^{-1}$ and t is the time elapsed in seconds.

What is the exact acceleration of the particle after 4 seconds?

A $\frac{\pi}{2}\,\mathrm{m\,s}^{-2}$

B $\frac{\pi}{4}\,\mathrm{m\,s}^{-2}$

C $\frac{\pi}{8}\,\mathrm{m\,s}^{-2}$

D $\frac{1}{4}\,\mathrm{m\,s}^{-2}$

Question 3

Position vectors $\overrightarrow{OA}$, $\overrightarrow{OB}$ and $\overrightarrow{OC}$ are equal to $\begin{pmatrix} 1 \\ -2p \\ 3q \end{pmatrix}$, $\begin{pmatrix} -3 \\ 2 - 2p \\ 3q - 1 \end{pmatrix}$ and $\begin{pmatrix} -4 \\ 1 - 2p \\ 3q - 4 \end{pmatrix}$, respectively, and p and q are real numbers.

Find the size of the angle between $\overrightarrow{BA}$ and $\overrightarrow{BC}$, correct to the nearest degree.

A $29°$

B $42°$

C $71°$

D $109°$

Question 4

What is the Cartesian form of $\underset{\sim}{r} = \frac{1}{\cos t}\underset{\sim}{i} + \frac{\sin t}{\cos t}\underset{\sim}{j}$?

A $x^2 + y^2 = 1$

B $x^2 - y^2 = 1$

C $y^2 + x^2 = -1$

D $y^2 - x^2 = 1$

Question 5

Which integral is equivalent to $\int_0^{\frac{\pi}{4}} \dfrac{\sec^2 x}{(1 + \tan x)^2}\, dx$ after an appropriate substitution?

A $\int_0^{\frac{\pi}{4}} \dfrac{1}{u}\, du$

B $\int_1^2 \dfrac{u^2}{(1 + u)^3}\, du$

C $\int_1^2 \dfrac{1}{u^2}\, du$

D $\int_0^{\frac{\pi}{4}} \dfrac{1}{u^3}\, du$

Question 6

Find $\int xe^x\, dx$.

A $xe^x + e^x + c$

B $xe^x - e^x + c$

C $-xe^x + e^x + c$

D $-xe^x - e^x + c$

Question 7

If $\int_1^5 f(x)\, dx = 8$, find the value of $\int_1^5 f(6 - x)\, dx$.

A -8

B -4

C 4

D 8

Question 8

A box slides down a slope of 60°.

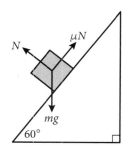

Given that the coefficient of friction is μ, what is the acceleration of the box?

A $\dfrac{1}{2}g\left(1 - \mu\sqrt{3}\right)$

B $\dfrac{1}{2}g\left(\mu\sqrt{3} - 1\right)$

C $\dfrac{1}{2}g\left(\sqrt{3} - \mu\right)$

D $\dfrac{1}{2}g\left(\mu - \sqrt{3}\right)$

Question 9

Which of the following is the contrapositive of $\neg P \Rightarrow \neg Q$?

A $\neg Q \Rightarrow P$

B $\neg P \Rightarrow Q$

C $Q \Rightarrow P$

D $P \Rightarrow Q$

Question 10

For 2 complex numbers a and b, which inequality is always true?

A $|a + b| \geq |a| + |b|$

B $|a - b| \geq |a| + |b|$

C $|a + b| \geq |a| - |b|$

D $|a - b| \leq |a| - |b|$

Section II

90 marks
Attempt Questions 11–16
Allow about 50 minutes for this section

- Answer the questions in the spaces provided. These spaces provide guidance for the expected length of response.
- Your responses should include relevant mathematical reasoning and/or calculations.

Question 11 (15 marks)

a   Given the complex numbers $z = 2 - 3i$ and $w = 3 - 2i$:

   i   express $z + w$ in polar form. 2 marks

   ii   express $\dfrac{z}{w}$ in the form $x + iy$, where x and y are real numbers. 2 marks

b   Sketch the region in the Argand diagram where $|z - 1| \le |z + 1|$ and $\dfrac{\pi}{3} \le \arg z \le \dfrac{2\pi}{3}$. 3 marks

Question 11 continues on page 88

Question 11 (continued)

c Evaluate $\int_0^1 (2x + 1) \sin(\pi x)\, dx$.   3 marks

d Find A, B and C such that $\dfrac{1}{x(x^2 + 9)} = \dfrac{A}{x} + \dfrac{Bx + C}{x^2 + 9}$.   2 marks

e Find $\displaystyle\int \dfrac{dx}{\sqrt{7 + 6x - x^2}}$.   3 marks

End of Question 11

Question 12 (15 marks)

a   Given $(k + 3)^3 = k^3 + 3k^2 + 3k + 1$, use mathematical induction to prove that   3 marks

$$\frac{1}{2} + \frac{2}{3} + \frac{3}{4} + \cdots + \frac{n}{n + 1} < \frac{n^2}{n + 1},$$

for all integers $n \geq 2$.

b   **i**   Given that $z = \cos\theta + i\sin\theta$, show that $\cos\theta = \frac{1}{2}\left(z + \frac{1}{z}\right)$.   1 mark

Question 12 continues on page 90

Question 12 (continued)

ii   Hence, write $\cos^5 \theta$ in terms of the cosine of multiples of θ.   3 marks

iii   Hence, find $\int \cos^5 \theta \, d\theta$.   2 marks

c   A particle is moving in a straight line and its position, x metres, from the origin, O, at time t seconds is given by the equation

$$x = 12 \cos 2t + 5 \sin 2t + 2.$$

i   Express $12 \cos 2t + 5 \sin 2t$ in the form $A \cos(2t - \alpha)$, where $0 < \alpha < \dfrac{\pi}{2}$ and $A > 0$.   2 marks

Question 12 continues on page 91

Question 12 (continued)

   ii   Prove that the particle is in simple harmonic motion, and find the amplitude of the motion.   2 marks

   iii   Find the maximum velocity of the particle and, correct to the nearest 0.01 second, when it first reaches this velocity.   2 marks

End of Question 12

Question 13 (15 marks)

a   Show that the line $r = \begin{pmatrix} 3 \\ 6 \\ 0 \end{pmatrix} + \mu \begin{pmatrix} -1 \\ 2 \\ -2 \end{pmatrix}$ is a tangent to the sphere given by $(x-2)^2 + (y-2)^2 + (z-1)^2 = 9$.   3 marks

b   *PQRS* is a trapezium.

Prove that $\overrightarrow{XY}$, which connects the midpoints of the parallel sides *PQ* and *SR*, is equal   3 marks
to half of $\left(\overrightarrow{PS} + \overrightarrow{QR}\right)$.

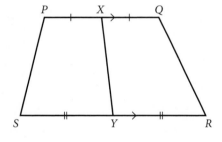

Question 13 continues on page 93

Question 13 (continued)

c   A sequence is given by the recurrence relation $a_1 = 7$, $a_{n+1} = 2a_n + 3$ for $n \geq 1$.   4 marks

Prove by mathematical induction that $a_n = 5(2^n) - 3$.

Question 13 continues on page 94

Question 13 (continued)

d   ©NESA 2016 HSC EXAM, QUESTION 16(a)

The complex numbers $z = \cos\theta + i\sin\theta$ and $w = \cos\alpha + i\sin\alpha$, where $-\pi < \theta \le \pi$ and $-\pi < \alpha \le \pi$, satisfy

$$1 + z + w = 0.$$

i   By considering the real and imaginary parts of $1 + z + w$, or otherwise, show that 1, z and w 3 marks
form the vertices of an equilateral triangle in the Argand diagram.

Question 13 continues on page 95

Question 13 (continued)

ii   Hence, or otherwise, show that if the 3 non-zero complex numbers $2i$, z_1 and z_2 satisfy   2 marks

$$|2i| = |z_1| = |z_2| \quad \text{AND} \quad 2i + z_1 + z_2 = 0,$$

then they form the vertices of an equilateral triangle in the Argand diagram.

Question 14 (15 marks)

a   A line passes through $P(2, -1, 3)$ and $Q(1, 3, -5)$.

i   Show that the vector equation of the line is given by   2 marks

$$\underset{\sim}{r} = (2\underset{\sim}{i} - \underset{\sim}{j} + 3\underset{\sim}{k}) + \lambda_1(-\underset{\sim}{i} + 4\underset{\sim}{j} - 8\underset{\sim}{k}), \lambda_1 \in \mathbb{R}.$$

ii   Determine if the point $R(4, -9, -13)$ lies on the line, $\underset{\sim}{r}$.   1 mark

Question 14 continues on page 96

Question 14 (continued)

iii A second line has a Cartesian equation given by   3 marks

$$\frac{x-2}{1} = \frac{y+2}{-2} = \frac{z-1}{12}.$$

Determine where this line and $\underset{\sim}{r}$ intersect.

_____

_____

_____

_____

_____

_____

_____

_____

_____

_____

_____

_____

_____

_____

_____

_____

_____

b A particle of mass m is released from rest to fall under gravity in a medium where the resistive force is mkv^2.

i Show that the distance fallen, x, is given by   3 marks

$$x = \frac{1}{2k}\ln\left(\frac{g}{g - kv^2}\right).$$

_____

_____

_____

_____

_____

_____

_____

Question 14 continues on page 97

Question 14 (continued)

ii Show that the velocity in terms of x is given by   2 marks

$$v = \sqrt{\frac{g}{k}(1 - e^{-2kx})}.$$

iii How far has the particle fallen when the velocity has reached half of the terminal velocity?   2 marks

c Given that $-\frac{1}{2} + i\frac{\sqrt{3}}{2} = e^{a+ib}$, where a and b are real, $-\pi < b \leq \pi$, find the exact values of a and b.   2 marks

End of Question 14

Question 15 (15 marks)

a Prove that $\dfrac{1+\sqrt{3}}{2}$ is irrational.   3 marks

b ©NESA 2001 HSC EXAM, QUESTION 8(a)

i Show that $2ab \leq a^2 + b^2$ for all real numbers a and b.   3 marks

Hence, deduce that $3(ab + bc + ca) \leq (a + b + c)^2$ for all real numbers a, b and c.

Question 15 continues on page 99

Question 15 (continued)

ii   Suppose a, b and c are the sides of a triangle.   4 marks

Explain why $(b - c)^2 \leq a^2$.

Deduce that $(a + b + c)^2 \leq 4(ab + bc + ca)$.

Question 15 continues on page 100

ii   Suppose a, b and c are the sides of a triangle.   4 marks

Question 15 (continued)

c   A particle of unit mass is moving along in a straight line with a resistive force of

$$R = kv,$$

where v metres per second is the velocity ($v > 0$) and k is a positive constant.

Its initial velocity is v_0 metres per second.

   i   Show that $\dfrac{dv}{v} = -k\,dt$.   1 mark

   ii   Show that the displacement of the particle is $x = \dfrac{(v_0 - v)t}{\ln v_0 - \ln v}$.   4 marks

End of Question 15

Question 16 (15 marks)

a   Show that the square root of $-7 + 24i$ is $\pm(2 + i)$, then use it to find the 4th roots of $-7 + 24i$.   4 marks

Question 16 continues on page 102

Question 16 (continued)

b   ©NESA   2020 HSC EXAM, QUESTION 16(b)

Let $I_n = \int_0^{\frac{\pi}{2}} \sin^{2n+1}(2\theta)\, d\theta$, $n = 0, 1, \ldots$.

i   Prove that $I_n = \dfrac{2n}{2n+1} I_{n-1}$, $n \geq 1$.   3 marks

Question 16 continues on page 103

Question 16 (continued)

ii   Deduce that $I_n = \dfrac{2^{2n}(n!)^2}{(2n+1)!}$.

3 marks

Question 16 continues on page 104

Question 16 (continued)

iii Let $J_n = \int_0^1 x^n (1-x)^n \, dx$, $n = 0, 1, 2, \ldots$.

3 marks

Using the result of part **ii**, or otherwise, show that $J_n = \dfrac{(n!)^2}{(2n+1)!}$.

Question 16 continues on page 105

Question 16 (continued)

iv   Prove that $(2^n n!)^2 \leq (2n + 1)!$.

2 marks

END OF PAPER

iv   Prove that $(2^n n!)^2 \leq (2n + 1)!$.

2 marks

WORKED SOLUTIONS

Section I (1 mark each)

Question 1

B $z^4 - 16 = (z^2 - 4)(z^2 + 4)$
$$= (z + 2)(z - 2)(z + 2i)(z - 2i)$$

> You need to factorise over the complex numbers, so $z^2 + 4$ can be further factorised.

Question 2

D $v = 2\tan^{-1}\left(\dfrac{t}{4}\right)$

$$a = \frac{dv}{dt} = 2\left[\frac{1}{1 + \left(\dfrac{t}{4}\right)^2}\right]\frac{1}{4}$$

When $t = 4$, $a = \dfrac{1}{4}$ m s^{-2}.

> Using $\ddot{x} = \dfrac{dv}{dt}$, make sure that the derivative is complete.

Question 3

D Using the scalar product,

$$\cos\theta = \frac{\begin{pmatrix} 4 \\ -2 \\ 1 \end{pmatrix} \cdot \begin{pmatrix} -1 \\ -1 \\ -3 \end{pmatrix}}{\left|\begin{pmatrix} 4 \\ -2 \\ 1 \end{pmatrix}\right|\left|\begin{pmatrix} -1 \\ -1 \\ -3 \end{pmatrix}\right|}$$

$$= \frac{-4 + 2 - 3}{\sqrt{21}\sqrt{11}}$$

$$= \frac{-5}{\sqrt{231}}$$

So $\theta \approx 109°$.

> Use the scalar product; don't forget to take square roots. The angle between the vectors is obtuse.

Question 4

B $x = \dfrac{1}{\cos t} = \sec t$, $y = \dfrac{\sin t}{\cos t} = \tan t$

We know that $1 + \tan^2 t = \sec^2 t$, so we get:

$$1 + y^2 = x^2$$
$$x^2 - y^2 = 1$$

> This is an exercise in converting parametric equations for x and y into a Cartesian equation. You need to use the result $1 + \tan^2\theta = \sec^2\theta$ to eliminate the parameter t.

Question 5

C Putting $u = 1 + \tan x$, $du = \sec^2 x\, dx$.

When $x = 0$, $u = 1$ and $x = \dfrac{\pi}{4}$, $u = 2$.

$$\int_0^{\frac{\pi}{4}} \frac{\sec^2 x}{(1 + \tan x)^2}\, dx = \int_1^2 \frac{du}{u^2}$$

> Some trial-and-error may be needed to find the correct substitution and result.

Question 6

B Integration by parts:

$$u = x \qquad v' = e^x$$
$$u' = 1 \qquad v = e^x$$

$$\int xe^x\, dx = xe^x - \int e^x\, dx$$
$$= xe^x - e^x + c$$

> A standard integration by parts question.

Question 7

D Putting $u = 6 - x$, $du = -dx$.

When $x = 1$, $u = 5$ and when $x = 5$, $u = 1$.

$$\int_1^5 f(6 - x)\, dx = \int_5^1 -f(u)\, du$$
$$= \int_1^5 f(u)\, du$$
$$= \int_1^5 f(x)\, dx$$
$$= 8$$

> Don't forget to change limits as these are critical to eliminating the negative sign.

WORKED SOLUTIONS

Question 8

C Resolving forces.

Perpendicular to the slope:

$N = mg\cos 60°$

$N = \dfrac{mg}{2}$ [1]

Along the slope:

$m\ddot{x} = mg\sin 60° - \mu N$

$m\ddot{x} = m\dfrac{g\sqrt{3}}{2} - \mu N$ [2]

Substitute [1] into [2]:

$m\ddot{x} = m\dfrac{g\sqrt{3}}{2} - \mu\dfrac{mg}{2}$

$\ddot{x} = \dfrac{g\sqrt{3}}{2} - \mu\dfrac{g}{2}$

$\ddot{x} = \dfrac{1}{2}g\left(\sqrt{3} - \mu\right)$

> It is necessary to establish the value of N in terms of m and g. Resolving forces acting on the box along the plane and perpendicular to the plane is required.

Question 9

C The contrapositive of $\neg P \Rightarrow \neg Q$ is $Q \Rightarrow P$.

> You must know definitions of converse, contrapositive and negation of a proposition.

Question 10

C Let $a = 2 + i$ and $b = 1 - i$.

Testing each option and using a counterexample:

A $|a + b| \geq |a| + |b|$ False

B $|a - b| \geq |a| + |b|$ False

C $|a + b| \geq |a| - |b|$ True

D $|a - b| \leq |a| - |b|$ False

> Using disproof by counterexample. This may take a couple of trials depending on choices for a and b. Don't confuse this with the triangle inequality.

Section II ($\checkmark$ = 1 mark)

Question 11

a i $z + w = 2 - 3i + 3 - 2i$

$$= 5 - 5i$$

$$|z + w| = \sqrt{5^2 + (-5)^2} = \sqrt{50} = 5\sqrt{2}$$

$$\arg(z + w) = \tan^{-1}\left(-\frac{5}{5}\right)$$

$$= -\frac{\pi}{4} \quad \checkmark \quad \text{(4th quadrant)}$$

$$z + w = 5\sqrt{2}\left[\cos\left(-\frac{\pi}{4}\right) + i\sin\left(-\frac{\pi}{4}\right)\right] \checkmark$$

Use definitions to find modulus and argument.

ii $\dfrac{z}{w} = \dfrac{2 - 3i}{3 - 2i} \times \dfrac{3 + 2i}{3 + 2i} \quad \checkmark$

$$= \frac{12 - 5i}{13}$$

$$= \frac{12}{13} - i\frac{5}{13} \quad \checkmark$$

Multiply by complex conjugate on top and bottom. Make sure that you express the answer in the required format, $x + iy$.

b $|z - 1| \le |z + 1|$ and $\dfrac{\pi}{3} \le \arg z \le \dfrac{2\pi}{3}$.

Let $z = x + iy$.

$$\sqrt{(x - 1)^2 + y^2} \le \sqrt{(x + 1)^2 + y^2}$$

$$(x - 1)^2 + y^2 \le (x + 1)^2 + y^2$$

$$x^2 - 2x + 1 \le x^2 + 2x + 1$$

$$-2x \le 2x$$

$$-4x \le 0$$

$$x \ge 0$$

So $x \ge 0$ $\checkmark$ and $\arg z$ lies between $60°$ and $120°$. $\checkmark$

Solve $|z - 1| \le |z + 1|$ after substituting $z = x + iy$. It should not be surprising that the solution is $x \ge 0$ because $|z - 1| \le |z + 1|$ means the distance of z from 1 is less than or equal to its distance from -1. Make sure that the lines are included and the origin is not included because arg 0 is undefined.

c $\int_0^1 (2x + 1)\sin(\pi x)\, dx$, using integration by parts:

$u = 2x + 1$ $v' = \sin(\pi x)$

$u' = 2$ $v = -\dfrac{1}{\pi}\cos(\pi x)$ $\checkmark$

$$= \left[(2x + 1)\left(-\frac{1}{\pi}\cos(\pi x)\right)\right]_0^1 + \frac{2}{\pi}\int_0^1 \cos(\pi x)\, dx \quad \checkmark$$

$$= \left(\frac{3}{\pi} + \frac{1}{\pi}\right) + \frac{2}{\pi^2}\left[\frac{1}{\pi}\sin(\pi x)\right]_0^1$$

$$= \frac{4}{\pi} + 0$$

$$= \frac{4}{\pi} \quad \checkmark$$

Using integration by parts, take care with negative signs.

d $\dfrac{1}{x(x^2 + 9)} = \dfrac{A}{x} + \dfrac{Bx + C}{x^2 + 9}$

$$\therefore 1 = A(x^2 + 9) + x(Bx + C)$$

$$= Ax^2 + 9A + Bx^2 + Cx \quad \checkmark$$

$$= (A + B)x^2 + Cx + 9A$$

$$\therefore A + B = 0 \qquad C = 0 \qquad 9A = 1$$

$$\therefore A = \frac{1}{9}, B = -\frac{1}{9} \text{ and } C = 0 \quad \checkmark$$

e $\int \dfrac{dx}{\sqrt{7 + 6x - x^2}} = \int \dfrac{dx}{\sqrt{16 - (9 - 6x + x^2)}}$ ✓

$$= \int \dfrac{dx}{\sqrt{4^2 - (x - 3)^2}}$$ ✓

$$= \sin^{-1}\left(\dfrac{x - 3}{4}\right) + c$$ ✓

Complete the square in the denominator (under the square root). Check your result by expanding and adjusting, if necessary. Know the standard integrals involving inverse trigonometric functions (on the HSC exam reference sheet).

Question 12

a $\dfrac{1}{2} + \dfrac{2}{3} + \dfrac{3}{4} + \cdots + \dfrac{n}{n + 1} < \dfrac{n^2}{n + 1}$

Prove true for $n = 2$:

LHS $= \dfrac{1}{2} + \dfrac{2}{3} = \dfrac{7}{6}$, RHS $= \dfrac{4}{3} = \dfrac{8}{6}$

LHS < RHS.

Hence, statement is true for $n = 2$. ✓

Assume true for $n = k$:

$\dfrac{1}{2} + \dfrac{2}{3} + \dfrac{3}{4} + \cdots + \dfrac{k}{k + 1} < \dfrac{k^2}{k + 1}$ ✓ [*]

Show true for $n = k + 1$:

RTP: $\dfrac{1}{2} + \dfrac{2}{3} + \dfrac{3}{4} + \cdots + \dfrac{k + 1}{k + 2} < \dfrac{(k + 1)^2}{k + 2}$

$$S_{k+1} = \left(\dfrac{1}{2} + \dfrac{2}{3} + \dfrac{3}{4} + \cdots + \dfrac{k}{k + 1}\right) + \dfrac{k + 1}{k + 2}$$

$$S_{k+1} < \dfrac{k^2}{k + 1} + \dfrac{k + 1}{k + 2} \quad \text{using assumption [*]}$$

$$= \dfrac{k^2(k + 2) + (k + 1)^2}{(k + 1)(k + 2)}$$

$$= \dfrac{k^3 + 3k^2 + 2k + 1}{(k + 1)(k + 2)}$$

$$< \dfrac{k^3 + 3k^2 + \mathbf{3}k + 1}{(k + 1)(k + 2)} \quad \text{as } k > 0 \ \ [\#]$$

$$= \dfrac{(k + 1)^3}{(k + 1)(k + 2)}$$

$$= \dfrac{(k + 1)^2}{(k + 2)}$$

$\therefore S_{k+1} < \dfrac{(k + 1)^2}{(k + 2)}$ ✓

Hence, if statement is true for $n = k$, then it is true for $n = k + 1$.

Since true for $n = 2$, then by mathematical induction it is true for all integers $n \geq 2$.

Check for $n = 2$ as the first case. State the assumption ($n = k$) and what is required to be shown for $n = k + 1$. Note the change to $3k$ in the numerator [#] which still keeps the inequality true and allows the result to be true.

b **i** $\dfrac{1}{2}\left(z + \dfrac{1}{z}\right) = \dfrac{1}{2}\left[(\cos\theta + i\sin\theta) + (\cos\theta - i\sin\theta)\right]$

$\qquad\qquad = \dfrac{1}{2}(2\cos\theta)$

$\qquad\qquad = \cos\theta$, as required. ✓

Remember, $\dfrac{1}{z} = z^{-1}$, which by De Moivre's theorem is $\cos(-\theta) + i\sin(-\theta) = \cos\theta - i\sin\theta$

ii $\cos^5\theta = \left[\dfrac{1}{2}\left(z + \dfrac{1}{z}\right)\right]^5$

$\qquad = \dfrac{1}{32}\left(z^5 + 5z^3 + 10z + \dfrac{10}{z} + \dfrac{5}{z^3} + \dfrac{1}{z^5}\right)$ (by binomial expansion) ✓

$\qquad = \dfrac{1}{32}\left[\left(z^5 + \dfrac{1}{z^5}\right) + 5\left(z^3 + \dfrac{1}{z^3}\right) + 10\left(z + \dfrac{1}{z}\right)\right]$ ✓

$\qquad = \dfrac{1}{32}\left[(2\cos5\theta) + 5(2\cos3\theta) + 10(2\cos\theta)\right]$ (by part **i**)

$\qquad = \dfrac{1}{16}(\cos5\theta + 5\cos3\theta + 10\cos\theta)$ ✓

A common HSC exam question. Use the result from part **i** after collecting together the terms in z^n.

iii $\displaystyle\int \cos^5\theta \, d\theta = \int \dfrac{1}{16}(\cos5\theta + 5\cos3\theta + 10\cos\theta) \, d\theta$ ✓

$\qquad\qquad = \dfrac{1}{16}\left(\dfrac{1}{5}\sin5\theta + \dfrac{5}{3}\sin3\theta + 10\sin\theta\right) + c$

$\qquad\qquad = \dfrac{1}{80}\sin5\theta + \dfrac{5}{48}\sin3\theta + \dfrac{5}{8}\sin\theta + c$ ✓

Replace $\cos^5\theta$ using the result in part **ii** above. Integrate and simplify where possible.

c **i** Let $12\cos2t + 5\sin2t = A\cos(2t - \alpha)$

$\qquad\qquad\qquad\qquad\quad = A(\cos2t\cos\alpha + \sin2t\sin\alpha)$

$\qquad\qquad\qquad\qquad\quad = A\cos2t\cos\alpha + A\sin2t\sin\alpha$

$\therefore A\cos\alpha = 12$ [1]

$\therefore A\sin\alpha = 5$ [2]

[2] ÷ [1]: $\qquad \tan\alpha = \dfrac{5}{12}$

$\qquad\qquad\qquad \alpha = \tan^{-1}\left(\dfrac{5}{12}\right)$ ✓

$[1]^2 + [2]^2$: $\qquad A^2\cos^2\alpha + A^2\sin^2\alpha = 5^2 + 12^2$

$\qquad\qquad\qquad A^2(\cos^2\alpha + \sin^2\alpha) = 169$

$\qquad\qquad\qquad\qquad\qquad\qquad A = \sqrt{169}$

$\qquad\qquad\qquad\qquad\qquad\qquad\quad = 13$

$\therefore 12\cos2t + 5\sin2t = 13\cos\left[2t - \tan^{-1}\left(\dfrac{5}{12}\right)\right]$ ✓

Use the result $\cos(A - B) = \cos A\cos B + \sin A\sin B$ for the auxiliary angle method.

ii $x = 12\cos 2t + 5\sin 2t + 2$

$$x = 13\cos\left[2t - \tan^{-1}\left(\frac{5}{12}\right)\right] + 2,$$

hence, the amplitude is 13. ✓

$$\dot{x} = -26\sin\left[2t - \tan^{-1}\left(\frac{5}{12}\right)\right]$$

$$\ddot{x} = -52\cos\left[2t - \tan^{-1}\left(\frac{5}{12}\right)\right]$$

$$= -4\left\{13\cos\left[2t - \tan^{-1}\left(\frac{5}{12}\right)\right]\right\}$$

$$= -4(x - 2) ✓$$

Hence, particle describes simple harmonic motion with amplitude of 13 and centre 2.

> Using the simplification from part **i**, rewrite x and find $\dot{x}$ and $\ddot{x}$. To show SHM, the equation of motion is either $\ddot{x} = -n^2 x$ or $\ddot{x} = -n^2(x - c)$.

iii $\dot{x} = -26\sin\left[2t - \tan^{-1}\left(\frac{5}{12}\right)\right]$

has the range $[-26, 26]$,
so maximum velocity = $26\,\mathrm{m\,s}^{-1}$. ✓

Find the time t when this first happens:

$$26 = -26\sin\left[2t - \tan^{-1}\left(\frac{5}{12}\right)\right]$$

$$-1 = \sin\left[2t - \tan^{-1}\left(\frac{5}{12}\right)\right]$$

$$2t - \tan^{-1}\left(\frac{5}{12}\right) = \frac{3\pi}{2}$$

$$\therefore\ t = \frac{1}{2}\left[\frac{3\pi}{2} + \tan^{-1}\left(\frac{5}{12}\right)\right]$$

$$\approx 2.55\text{ seconds ✓}$$

> Notice that the applied problems in SHM use a variety of techniques and facts: the range of the sine function, exact values, trigonometric equations, inverse tangent function.
>
> An easy mistake is to solve for the first time where $\ddot{x} = 0$. But here, that approach gives a minimum (negative) velocity. Solving for $v = 26$ avoids this problem.

Question 13

a We find the point of intersection by substituting the line components into the sphere equation:

$$(3 - \lambda - 2)^2 + (6 + 2\lambda - 2)^2 + (-2\lambda - 1)^2 = 9$$
$$9\lambda^2 + 18\lambda + 18 = 9$$

which gives $\lambda = -1$. Substituting λ into $\underset{\sim}{r}$ gives us our point of intersection $P(4, 4, 2)$. ✓

Let $C(2, 2, 1)$ represent the centre of the sphere. To show that $\underset{\sim}{r}$ is a tangent, its direction vector must be perpendicular to $\overrightarrow{CP}$. ✓

$$\begin{pmatrix} -1 \\ 2 \\ -2 \end{pmatrix} \cdot \overrightarrow{CP} = \begin{pmatrix} -1 \\ 2 \\ -2 \end{pmatrix} \cdot \begin{pmatrix} 2 \\ 2 \\ 1 \end{pmatrix} ✓$$

$$= -2 + 4 - 2$$
$$= 0,\text{ as required.}$$

Alternatively, we may note that there is only one solution for λ, implying that the line touches the sphere at one and only one point, which also implies that it is a tangent.

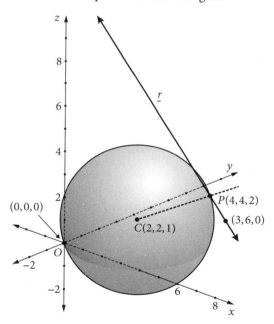

Quite a complex 3D geometry problem.

b

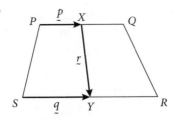

$$\overrightarrow{PS} = \underset{\sim}{p} + \underset{\sim}{r} - \underset{\sim}{q} \checkmark$$

$$\overrightarrow{QR} = -\underset{\sim}{p} + \underset{\sim}{r} + \underset{\sim}{q}$$

$$\overrightarrow{PS} + \overrightarrow{QR} = \underset{\sim}{p} + \underset{\sim}{r} - \underset{\sim}{q} - \underset{\sim}{p} + \underset{\sim}{r} + \underset{\sim}{q}$$

$$PS = 2\underset{\sim}{r} \checkmark$$

$$= 2\overrightarrow{XY}, \text{ as required. } \checkmark$$

c For $n = 1$:

$a_1 = 7$ and $a_1 = 5(2^1) - 3 = 7$

Therefore, statement is true for $n = 1$. ✓

Assume statement is true for $n = k$:

$a_k = 5(2^k) - 3$ and $a_{k+1} = 2a_k + 3$. [*] ✓

For $n = k + 1$, show that $a_{k+1} = 5(2^{k+1}) - 3$.

Now, $a_{k+1} = 2a_k + 3$

$$= 2[5(2^k) - 3] + 3 \quad \text{by assumption } [*]$$

$$= 5(2^{k+1}) - 6 + 3$$

$$= 5(2^{k+1}) - 3, \text{ as required. } \checkmark$$

If true for $n = k$, then true for $n = k + 1$.

It is true for $n = 1$, so by mathematical induction it is true for all integers $n \geq 1$. ✓

> It helps to write down the expression to be shown for a_{k+1}. Make sure you always complete the process of using mathematical induction.

d **i** $z = \cos\theta + i\sin\theta = 1e^{i\theta}$

$w = \cos\alpha + i\sin\alpha = 1e^{i\alpha}$

Using $1 + z + w = 0$:

Real components:

$1 + \cos\theta + \cos\alpha = 0$ [1]

Imaginary components:

$\sin\theta + \sin\alpha = 0$ [2] ✓

$|z| = |w| = 1$,
so z and w lie on the unit circle.

From [2]:

$\sin\theta = -\sin\alpha$

$\therefore \theta = -\alpha$, or $\alpha + \pi$, or $\alpha - \pi$

If $\theta = \alpha \pm \pi$, then $\cos\theta = -\cos\alpha$

But from [1] we see that $\cos\theta + \cos\alpha = -1$, so this is not possible.

$\therefore \theta = -\alpha$

Substitute into [1]:

$1 + \cos(-\alpha) + \cos\alpha = 0$

$$1 + 2\cos\alpha = 0$$

$$\cos\alpha = -\frac{1}{2}$$

$$\alpha = -\frac{2\pi}{3} \text{ and } \theta = \frac{2\pi}{3}$$

or vice versa ✓

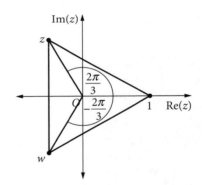

1, z and w lie on the unit circle and are equally spaced around the origin $\left(\frac{2\pi}{3}\right)$, hence they form the vertices of an equilateral triangle.

> Use a diagram whenever you can to get further insight into the question. This 2016 HSC exam question tested visual understanding, logic and the ability to explain in words.

ii $|2i| = |z_1| = |z_2|$ means that $2i$, z_1 and z_2 are all located on a circle of radius 2. ✓

Also, $2i + z_1 + z_2 = 0$.

Divide both sides by $2i$:

$$1 + \frac{z_1}{2i} + \frac{z_2}{2i} = 0$$

Let $z = \frac{z_1}{2i}$ and $w = \frac{z_2}{2i}$.

Then $1 + z + w = 0$ and $|z| = |w| = 1$ as in part **i**.

∴ 1, z and w form the vertices of an equilateral triangle.

But $2i$, $2iz$ and $2iw$ are the above values multiplied by $2i$, respectively, which geometrically is doubling their moduli and rotating them $\frac{\pi}{2}$ anticlockwise, creating a similar triangle of scale factor 2.

This means $2i$, z_1 and z_2 form the vertices of an equilateral triangle. ✓

> Use the modulus and argument for each point to establish the required result. This was a challenging problem that appeared in the last question of the 2016 HSC exam.

Question 14

a i The direction vector
$$= (1 - 2, 3 - (-1), -5 - 3)$$
$$= (-1, 4, -8) ✓$$

Choosing the point $P(2, -1, 3)$ through which the line passes we get:

$$r = (2i - j + 3k) + \lambda_1(-i + 4j - 8k). ✓$$

> Use P and Q to obtain the direction vector. Choosing the correct point P will get the required result.

ii Substituting the point $R(4, -9, -13)$ into this line, we see that:

$$\begin{aligned} 4 &= 2 - \lambda_1 &\to \lambda_1 &= -2 \\ -9 &= -1 + 4\lambda_1 &\to \lambda_1 &= -2 \\ -13 &= 3 - 8\lambda_1 &\to \lambda_1 &= 2. \end{aligned}$$

Since these parameters are not all the same, then the point R does not lie on the line. ✓

> All parameters must be the same for R to lie on the line.

iii The given line is
$$r_1 = (2i - j + 3k) + \lambda_1(-i + 4j - 8k).$$

The second line is given by
$$r_2 = (2i - 2j + k) + \lambda_2(i - 2j + 12k). ✓$$

Solving simultaneously:

$$\begin{aligned} 2 + \lambda_2 &= 2 - \lambda_1 & [1] \\ -2 - 2\lambda_2 &= -1 + 4\lambda_1 & [2] \\ 1 + 12\lambda_2 &= 3 - 8\lambda_1 & [3] \end{aligned}$$

From [1]:

$$\lambda_2 = -\lambda_1$$

Substitute into [2]:

$$-2 - 2(-\lambda_1) = -1 + 4\lambda_1$$
$$-2\lambda_1 = 1$$

$$\lambda_1 = -\frac{1}{2} \quad \text{and} \quad \lambda_2 = \frac{1}{2} ✓$$

Checking for consistency in [3]:

$$\text{LHS} = 1 + 12\left(-\frac{1}{2}\right) = 7$$

$$\text{RHS} = 3 - 8\left(-\frac{1}{2}\right) = 7 \qquad \text{LHS} = \text{RHS}$$

Hence, the lines meet at $\left(2\frac{1}{2}, -3, 7\right)$. ✓

> Equating components will give a series of simultaneous equations. You need only solve 2 equations simultaneously but then it is necessary to check the third equation to ensure that all components are satisfying both lines. Take care with negative numbers when solving equations.

b i $m\ddot{x} = mg - mkv^2$ ✓

$$v\frac{dv}{dx} = g - kv^2$$

$$\int \frac{v}{g - kv^2} dv = \int dx$$

$$-\frac{1}{2k}\ln(g - kv^2) = x + c ✓$$

Note: at $t = 0$, $v = 0$ and $g - kv^2 > 0$, so we can drop the absolute-value signs when integrating.

When $x = 0$, $v = 0$, ∴ $c = -\frac{1}{2k}\ln g$.

Hence, $x = \frac{1}{2k}\ln g - \frac{1}{2k}\ln(g - kv^2)$ $\ddot{x} = g - kv^2 > 0$

$$x = \frac{1}{2k}\ln\left(\frac{g}{g - kv^2}\right), \text{ as required. } ✓$$

> A diagram will assist you in determining the direction of forces. As the particle is released from rest, then when $t = 0$, $v = 0$.

WORKED SOLUTIONS

ii Rearranging previous answer:

$$2kx = \ln\left(\frac{g}{g - kv^2}\right) \checkmark$$

$$2kx = -\ln\left(\frac{g - kv^2}{g}\right)$$

$$ge^{-2kx} = g - kv^2$$
$$kv^2 = g - ge^{-2kx}$$

$$\therefore v = \sqrt{\frac{g}{k}(1 - e^{-2kx})}, \text{ as required. } \checkmark$$

Note that $v \geq 0$ for $t \geq 0$.

iii Find x when $v = \frac{1}{2}v_T$.

Terminal velocity as $x \to \infty$, $v_T = \sqrt{\frac{g}{k}}$ $\checkmark$

$$\therefore \frac{1}{2}v_T = \frac{1}{2}\sqrt{\frac{g}{k}}$$

so $v^2 = \frac{1}{4}\left(\frac{g}{k}\right)$.

Using $x = \frac{1}{2k}\ln\left(\frac{g}{g - kv^2}\right)$,

$$x = \frac{1}{2k}\ln\left(\frac{g}{g - k\left(\frac{g}{4k}\right)}\right)$$

$$= \frac{1}{2k}\left(\frac{1}{1 - \frac{1}{4}}\right)$$

$$= \frac{1}{2k}\ln\left(\frac{4}{3}\right) \checkmark$$

Using $\ddot{x} = 0$ or $1 - e^{-2kx} \to 0$ as $x \to \infty$ to find v_T.
Note also that it is easier to substitute for v^2 into the expression for x.

c Write $-\frac{1}{2} + i\frac{\sqrt{3}}{2}$ in exponential form:

$$r = \sqrt{\left(-\frac{1}{2}\right)^2 + \left(\frac{\sqrt{3}}{2}\right)^2}$$

$$= \sqrt{\frac{1}{4} + \frac{3}{4}}$$

$$= \sqrt{1}$$

$$= 1$$

$$\tan\theta = \frac{\sqrt{3}}{2} \div \left(-\frac{1}{2}\right)$$

$$= -\sqrt{3}$$

$$\theta = \frac{2\pi}{3} \checkmark$$

$$\therefore -\frac{1}{2} + i\frac{\sqrt{3}}{2} = 1e^{\frac{2\pi i}{3}} = e^{a+ib}$$

so equating real and imaginary,

we get $a = 0$ and $b = \frac{2\pi}{3}$. $\checkmark$

Be careful – you are after values for a and b.
Note: Arg z is in the 2nd quadrant.

WORKED SOLUTIONS

Question 15

a Prove by contradiction that $\dfrac{1+\sqrt{3}}{2}$ is irrational.

Assume $\sqrt{3}$ is rational, $\sqrt{3} = \dfrac{p}{q}$, where p, q are integers which have no common factor and $q \neq 0$.

$\therefore 3 = \dfrac{p^2}{q^2}$ and so $3q^2 = p^2$. ✓

So p^2 is divisible by 3.

Since 3 is prime, it follows that p is divisible by 3,

so $p = 3k$, $k \in \mathbb{Z}$.

$\therefore 3q^2 = (3k)^2$ ✓
$\quad q^2 = 3k^2$

So q^2 is divisible by 3, which implies that q is also divisible by 3.

This is a contradiction since p, q have no common factor.

Therefore, $\sqrt{3}$ is irrational and so $\dfrac{1+\sqrt{3}}{2}$ is irrational as well. ✓

A common exam question so learn this proof.

Proving $\dfrac{1+\sqrt{3}}{2}$ is irrational is really about showing $\sqrt{3}$ is irrational.

b i $(a-b)^2 \geq 0$

$\therefore a^2 - 2ab + b^2 \geq 0$

$\therefore a^2 + b^2 \geq 2ab$

> **Hint**
> This proof can also be started by considering the difference $a^2 + b^2 - 2ab$.

Thus, $2ab \leq a^2 + b^2$, as required. ✓

Similarly, $2ac \leq a^2 + c^2$, $2bc \leq b^2 + c^2$. ✓

So adding all 3 inequalities together, we get:

$2ab + 2ac + 2bc \leq 2a^2 + 2b^2 + 2c^2$
$\quad ab + ac + bc \leq a^2 + b^2 + c^2$

Adding $2ab + 2bc + 2ac$ to both sides:

$ab + ac + bc + 2ab + 2bc + 2ac \leq a^2 + b^2 + c^2 + 2ab + 2bc + 2ac$
$\qquad\qquad 3ab + 3bc + 3ac \leq a^2 + b^2 + c^2 + 2ab + 2bc + 2ac$
$\qquad\qquad\quad 3(ab + bc + ac) \leq (a + b + c)^2$, as required. ✓

A fairly challenging HSC exam question, especially with the second half. Note: Any number squared is greater or equal to 0. There are similar results to $2ab$ with numbers a with c, and b with c. These are then combined to lead to the required result.

9780170459273

ii By the triangle inequality:

$$a + b \geq c$$
$$a \geq c - b \text{ and similarly } a \geq b - c$$
$$|a| \geq |c - b|$$
$$a^2 \geq (c - b)^2$$
$$a^2 \geq (b - c)^2$$

$(b - c)^2 \leq a^2$, as required.

OR using the cosine rule:

$$c^2 = a^2 + b^2 - 2ab \cos C$$
$$\cos C = \frac{a^2 + b^2 - c^2}{2ab}$$

Since $\cos C \leq 1$,

$$\frac{a^2 + b^2 - c^2}{2ab} \leq 1$$
$$a^2 + b^2 - c^2 \leq 2ab \quad (2ab > 0)$$
$$a^2 + b^2 - 2ab \leq c^2$$
$$(a - b)^2 \leq c^2$$

Similarly,

$(b - c)^2 \leq a^2$, as required. ✓

Similarly,

since $(a - b)^2 \leq c^2$,

then $(a - c)^2 \leq b^2$.

Adding all 3 inequalities:

$$(b - c)^2 + (a - b)^2 + (a - c)^2 \leq a^2 + c^2 + b^2$$
$$b^2 - 2bc + c^2 + a^2 - 2ab + b^2 + a^2 - 2ac + c^2 \leq a^2 + b^2 + c^2$$
$$a^2 + b^2 + c^2 - 2bc - 2ab - 2ac \leq 0 \quad ✓ \quad [*]$$

But $(a + b + c)^2 = a^2 + ab + ac + ba + b^2 + bc + ca + cb + c^2$
$$= a^2 + 2ab + 2ac + b^2 + 2bc + c^2.$$

$\therefore a^2 + b^2 + c^2 = (a + b + c)^2 - 2ab - 2ac - 2bc$ ✓

So substituting into [*]:

$$(a + b + c)^2 - 2ab - 2ac - 2bc - 2bc - 2ab - 2ac \leq 0$$
$$(a + b + c)^2 - 4ab - 4ac - 4bc \leq 0$$
$$(a + b + c)^2 \leq 4ab + 4ac + 4bc$$
$$(a + b + c)^2 \leq 4(ab + bc + ca), \text{ as required. } ✓$$

This was from part **a** of the last question in the 2001 HSC exam, a fairly difficult question with which some students struggled.

A triangle with sides a, b and c might imply using the triangle inequality or cosine rule. Note the symmetry in the inequalities for a, b, c. Watch out for the inequalities that may change with any rearrangement of terms.

9780170459273

c i $m\ddot{x} = -mkv$

$$\frac{dv}{dt} = -kv$$

$$\frac{dv}{v} = -k\,dt \quad \checkmark$$

Unit mass implies $m = 1$.

ii Integrating

$\ln v = -kt + c \quad (v > 0) \quad \checkmark$

When $t = 0$, $v = v_0$.

$\ln v_0 = 0 + c$

$\quad c = \ln v_0$

$\therefore \ln v = -kt + \ln v_0$

$\quad kt = \ln v_0 - \ln v$

$\quad t = \dfrac{1}{k}\ln\left(\dfrac{v_0}{v}\right) \quad [*] \quad \checkmark$

Find v as a function of x:

$v\dfrac{dv}{dx} = -kv$

$\dfrac{dv}{dx} = -k$

$\quad v = -kx + d$

When $x = 0$, $v = v_0$:

$v_0 = 0 + d$

$\quad v = -kx + v_0 \quad \checkmark$

$\quad kx = v_0 - v$

$\quad x = \dfrac{v_0 - v}{k}$

From $[*]$, $\dfrac{1}{k} = \dfrac{t}{\ln\left(\dfrac{v_0}{v}\right)}$

$\therefore x = \dfrac{(v_0 - v)t}{\ln\left(\dfrac{v_0}{v}\right)} = \dfrac{(v_0 - v)t}{\ln v_0 - \ln v}$, as required. $\checkmark$

Note that the final expression for x has no k. It is best to eliminate k by using $[*]$. The displacement, x, can be expressed in terms of only v, v_0, and t.

Question 16

a Let $\sqrt{-7 + 24i} = x + iy \quad (x, y \text{ real})$

$\quad -7 + 24i = x^2 + 2xyi - y^2$

$\therefore x^2 - y^2 = -7$ and $2xy = 24$

$$y = \frac{12}{x} \quad \checkmark$$

Solving:

$$x^2 - \left(\frac{12}{x}\right)^2 = -7$$

$$x^2 - \frac{144}{x^2} = -7$$

$$x^4 + 7x^2 - 144 = 0$$

$$(x^2 + 16)(x^2 - 9) = 0$$

$\therefore x = \pm 3 \quad \therefore y = \pm 4$

So $z^2 = \pm\sqrt{-7 + 24i} = \pm(3 + 4i)$. $\checkmark$

Repeating the process to solve $z^4 = -7 + 24i$ to find the 4th roots of $-7 + 24i$:

$z = \pm\sqrt{3 + 4i} \quad \text{OR} \quad \pm\sqrt{-3 - 4i}$

Let $z = x + iy$:

Case 1: $x^2 - y^2 = 3$ and $2xy = 4$

$$y = \frac{2}{x}$$

Solving:

$$x^2 - \left(\frac{2}{x}\right)^2 = 3$$

$$x^2 - \frac{4}{x^2} = 3$$

$$x^4 - 3x^2 - 4 = 0$$

$$(x^2 - 4)(x^2 + 1) = 0$$

$\therefore x = \pm 2 \quad \therefore y = \pm 1 \quad \therefore z = \pm(2 + i) \quad \checkmark$

Case 2: $x^2 - y^2 = -3$ and $2xy = -4$

$$y = \frac{-2}{x}$$

Solving:

$$x^2 - \left(\frac{-2}{x}\right)^2 = -3$$

$$x^2 - \frac{4}{x^2} = -3$$

$$x^4 + 3x^2 - 4 = 0$$

$$(x^2 + 4)(x^2 - 1) = 0$$

$\therefore x = \pm 1 \quad \therefore y = \mp 2 \quad \therefore z = \pm(1 - 2i)$

Solutions are $2 + i$, $-2 - i$, $1 - 2i$, $-1 + 2i$. $\checkmark$

There are 2 expressions for z^2.
There are 4 expressions for z.

Be careful with pairing of signs and noting there are 4 distinct solutions.

WORKED SOLUTIONS

b i $I_n = \int_0^{\frac{\pi}{2}} \sin^{2n+1}(2\theta)\, d\theta$

$I_n = \int_0^{\frac{\pi}{2}} \sin^{2n}(2\theta) \sin(2\theta)\, d\theta$

Using integration by parts:

$u = \sin^{2n} 2\theta,$ $\qquad\qquad v' = \sin 2\theta$

$u' = 4n \sin^{2n-1} 2\theta \cos 2\theta,$ $\qquad v = -\frac{1}{2}\cos 2\theta$ ✓

$$I_n = \left[-\frac{1}{2}\cos 2\theta \sin^{2n} 2\theta \right]_0^{\frac{\pi}{2}} + 2n \int_0^{\frac{\pi}{2}} \sin^{2n-1} 2\theta \times \cos^2 2\theta\, d\theta$$

$$= [0 - 0] + 2n \int_0^{\frac{\pi}{2}} \sin^{2n-1} 2\theta\, (1 - \sin^2 2\theta)\, d\theta$$

$$= 2n \int_0^{\frac{\pi}{2}} \sin^{2n-1} 2\theta - \sin^{2n+1} 2\theta\, d\theta$$

$$= 2n \int_0^{\frac{\pi}{2}} \sin^{2(n-1)+1} 2\theta - \sin^{2n+1} 2\theta\, d\theta \quad ✓$$

$$= 2n\left[I_{n-1} - I_n \right]$$

$$I_n = 2nI_{n-1} - 2nI_n$$

$$I_n + 2nI_n = 2nI_{n-1}$$

$$I_n(1 + 2n) = 2nI_{n-1}$$

$$\therefore I_n = \frac{2n}{2n+1} I_{n-1}, \text{ as required. } ✓$$

These challenging proofs and integrals made up the very last question of the 2020 exam. It requires much practice or trial-and-error to break these trigonometric expressions apart to obtain the desired result. The hint is usually in the answer, in this case you need to get I_{n-1}. This is not always easy to do. When setting out a proof, it is important to show your working clearly and not skip steps. For example, the 4 components of an integration by parts should be listed. Choosing the correct u and v' is crucial.

ii $\therefore I_n = \dfrac{2n}{2n+1} I_{n-1}$

$= \dfrac{2n}{2n+1}\left[\dfrac{2(n-1)}{2(n-1)+1}\right] I_{n-2}$

$= \dfrac{2n}{2n+1}\left[\dfrac{2n-2}{2n-1}\right]\left[\dfrac{2(n-2)}{2(n-2)+1}\right] I_{n-3}$

$= \dfrac{2n}{2n+1}\left[\dfrac{2n-2}{2n-1}\right]\left[\dfrac{2n-4}{2n-3}\right]\left[\dfrac{2n-6}{2n-5}\right] I_{n-4}$

$= \dfrac{2n}{2n+1}\left[\dfrac{2n-2}{2n-1}\right]\left[\dfrac{2n-4}{2n-3}\right]\left[\dfrac{2n-6}{2n-5}\right]\cdots\dfrac{2}{3} I_0$ ✓

But $I_0 = \displaystyle\int_0^{\frac{\pi}{2}} \sin 2\theta \, d\theta$

$= \left[-\dfrac{1}{2}\cos 2\theta\right]_0^{\frac{\pi}{2}}$

$= \left[-\dfrac{1}{2}\cos \pi - -\dfrac{1}{2}\cos 0\right]$

$= \left[-\dfrac{1}{2}(-1) + \dfrac{1}{2}(1)\right]$

$= 1$ ✓

$\therefore I_n = \dfrac{2n}{2n+1}\left[\dfrac{2n-2}{2n-1}\right]\left[\dfrac{2n-4}{2n-3}\right]\left[\dfrac{2n-6}{2n-5}\right]\cdots\dfrac{2}{3}$

$= \dfrac{2(n)2(n-1)2(n-2)2(n-3)\ldots2(1)}{(2n+1)(2n-1)(2n-3)(2n-5)\ldots3}$

$= \dfrac{2^n(n!)}{(2n+1)(2n-1)(2n-3)(2n-5)\ldots3}$

To make denominator $(2n+1)!$, multiply numerator and denominator by the product of 'even' terms: $2n(2n-2)(2n-4)\ldots2$

$I_n = \dfrac{2^n(n!)}{(2n+1)(2n-1)(2n-3)(2n-5)\ldots3} \times \dfrac{2n(2n-2)(2n-4)(2n-6)\ldots2}{2n(2n-2)(2n-4)(2n-6)\ldots2}$

$= \dfrac{2^n(n!)}{(2n+1)!} \times \dfrac{2n(2[n-1])(2[n-2])(2[n-3])\ldots2[1]}{1}$

$= \dfrac{2^n(n!)}{(2n+1)!} \times \dfrac{2^n(n!)}{1}$

$I_n = \dfrac{2^{2n}(n!)^2}{(2n+1)!}$ ✓

This recurrence relation integral is difficult and requires foresight and spotting a pattern. By substituting into the expression from part **i** repeatedly, you get close to the required result. Noting what is missing and determining how to go further based on this information is necessary. Many students forgot to find the value of I_0.

iii $J_n = \int_0^1 x^n (1-x)^n \, dx$

Putting $1 - x = \sin^2 \theta$

$$x = 1 - \sin^2 \theta$$
$$dx = -2 \sin \theta \cos \theta \, d\theta \quad \checkmark$$

When $x = 0$, $\theta = \dfrac{\pi}{2}$ and $x = 1$, $\theta = 0$:

Substituting:

$$J_n = \int_{\frac{\pi}{2}}^0 (1 - \sin^2 \theta)^n (\sin^2 \theta)^n \times -2 \sin \theta \cos \theta \, d\theta$$

$$= -2 \int_{\frac{\pi}{2}}^0 \cos^{2n} \theta \sin^{2n} \theta \sin \theta \cos \theta \, d\theta$$

Swapping the limits of integration:

$$J_n = 2 \int_0^{\frac{\pi}{2}} \sin^{2n+1} \theta \cos^{2n+1} \theta \, d\theta$$

$$= \frac{2}{2^{2n+1}} \int_0^{\frac{\pi}{2}} 2^{2n+1} \sin^{2n+1} \theta \cos^{2n+1} \theta \, d\theta$$

$$= \frac{1}{2^{2n}} \int_0^{\frac{\pi}{2}} \sin^{2n+1} 2\theta \, d\theta \quad \checkmark$$

Using part **ii**:

$$J_n = \frac{1}{2^{2n}} \frac{(2^n)^2 (n!)^2}{(2n+1)!}$$

$$= \frac{(n!)^2}{(2n+1)!}, \text{ as required.} \quad \checkmark$$

> Another complex proof requiring a diverse range of mathematical skills, including knowing to take out a factor of $\dfrac{1}{2^{2n}}$. As this is part **iii**, you could expect that it is connected to parts **i** and/or **ii**.
>
> It is necessary to find a connection between the given result for J_n and I_n in part **i**.
>
> The substitution $x = \sin^2 \theta$ or $x = \cos^2 \theta$ also works.
>
> Note that $\sin 2\theta = 2 \sin \theta \cos \theta$. Commit this trigonometric identity to memory as it is used often.

iv Now, $f(x) = x(1 - x)$ has a maximum value of $\dfrac{1}{4}$ when $x = \dfrac{1}{2}$.

So $x^n (1 - x)^n$ has a maximum value of

$$\left(\frac{1}{4}\right)^n = \frac{1}{2^{2n}}. \quad \checkmark$$

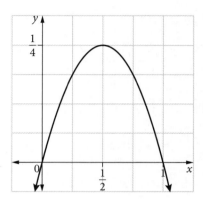

So the area under the curve $y = x^n (1 - x)^n$ will be less than the area of a rectangle with height $\dfrac{1}{2^{2n}}$ and width 1.

$$\therefore J_n \leq \int_0^1 \frac{1}{2^{2n}} \, dx$$

$$= \left[\frac{1}{2^{2n}} x \right]_0^1$$

$$= \frac{1}{2^{2n}} (1 - 0)$$

$$= \frac{1}{2^{2n}}$$

$$\therefore \frac{(n!)^2}{(2n+1)!} \leq \frac{1}{2^{2n}}$$

$$2^{2n} (n!)^2 \leq (2n+1)!$$

$$(2^n n!)^2 \leq (2n+1)! \text{ as required.} \quad \checkmark$$

> Again, observe the similarity in the part **iii** result and the inequality in the result to be proved in part **iv**. Noting that $x(1 - x) \leq \dfrac{1}{4}$, you can then obtain an expression for $x(1 - x)$ and finally $x^n (1 - x)^n$. This can be followed up by using J_n and the answer from part **iii** to get the required result.
>
> This question requires you to use logic and explain your reasoning. There are also many alternative solutions to this question, some more complex and time-consuming than others. Choose the simplest and most efficient method.

Mathematics Extension 2

PRACTICE HSC EXAM 2

General instructions	• Reading time: 10 minutes
	• Working time: 3 hours
	• A reference sheet is provided on page 179 at the back of this book
	• For questions in Section II, show relevant mathematical reasoning and/or calculations

Total marks: 100	**Section I – 10 questions, 10 marks**
	• Attempt Questions 1–10
	• Allow about 15 minutes for this section
	Section II – 6 questions, 90 marks
	• Attempt Questions 11–16
	• Allow about 2 hours and 45 minutes for this section

Section I

10 marks
Attempt Questions 1–10
Allow about 15 minutes for this section

Circle the correct answer.

Question 1

Find $\int \dfrac{4 - x}{(x - 1)(x + 2)} \, dx$.

A $\ln|x - 1| + 2\ln|x + 2| + c$

B $-\ln|x - 1| + 2\ln|x + 2| + c$

C $\ln|x - 1| - 2\ln|x + 2| + c$

D $-\ln|x - 1| - 2\ln|x + 2| + c$

Question 2

Which point below lies on the line with vector equation $\underset{\sim}{r} = \begin{pmatrix} 1 \\ -1 \\ 1 \end{pmatrix} + \lambda \begin{pmatrix} -2 \\ 1 \\ -1 \end{pmatrix}$?

A $(-2, 1, -1)$

B $(-3, 2, -2)$

C $(-1, 0, 0)$

D $(3, -2, 0)$

Question 3

Which diagram shows the solutions to the equation $z^5 + 1 = 0$ as vertices of a pentagon on the complex plane?

A
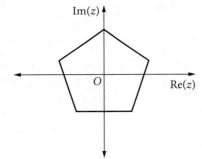

B
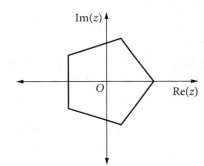

C
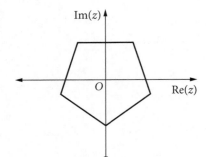

D
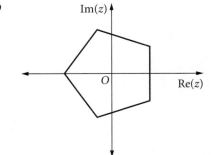

Question 4

A particle P moving about the origin O in simple harmonic motion is stationary at $x = 5$, then takes 3 seconds to travel to O.

What is the maximum speed of P?

A $\dfrac{\pi}{6}$

B $\dfrac{5\pi}{6}$

C $\dfrac{5\pi}{3}$

D $\dfrac{10\pi}{3}$

Question 5

What is the negation of this statement?

'All composite numbers have more than two factors.'

A   'All composite numbers have less than two factors.'

B   'At least one composite number does not have two factors.'

C   'All composite numbers have two factors.'

D   'At least one composite number does not have more than two factors.'

Question 6

The point $(5, -2, 7)$ is reflected in the x-z plane. What are the coordinates of the new point?

A $(-5, -2, -7)$

B $(5, 2, -7)$

C $(-5, 2, 7)$

D $(5, 2, 7)$

Question 7

Solve $\dfrac{x+1}{(x-1)(x+3)} \geq 0$.

A $x < -3$ or $-1 \leq x < 1$

B $-3 < x \leq -1$ or $x > 1$

C $x > 3$ or $-1 < x \leq 1$

D $x < -1$ or $1 \leq x < 3$

Question 8

For the differential equation $\ddot{x} = 4 - x$, what is a correct solution for $\dot{x}$?

A $\dot{x} = 4x - \dfrac{x^2}{2} + c$

B $\dot{x} = 4t - xt + c$

C $\dot{x} = \sqrt{8x - x^2 + c}$

D $\dot{x}\dfrac{dv}{dx} = 4v - x\dfrac{dv}{dx} + c$

Question 9

Evaluate $\int_0^2 \sqrt{x(2-x)}\,dx$.

A $\dfrac{\pi}{4}$

B $\dfrac{\pi}{2}$

C π

D 2π

Question 10

Consider two points, z_1 and z_2, on an Argand diagram.

If $\arg\left(\dfrac{z_1 + z_2}{z_2 - z_1}\right) = \dfrac{\pi}{2}$, then which of the following equations must be true?

A $|z_1| = |z_2|$

B $|z_1 + z_2| = |z_2 - z_1|$

C $\arg\left(\dfrac{z_1 + z_2}{z_1}\right) = \arg\left(\dfrac{z_1 + z_2}{z_2}\right)$

D $(z_2 - z_1)i = z_1 + z_2$

Section II

90 marks
Attempt Questions 11–16
Allow about 2 hours 45 minutes for this section

- Answer the questions in the spaces provided. These spaces provide guidance for the expected length of response.
- Your responses should include relevant mathematical reasoning and/or calculations.

Question 11 (15 marks)

a   **i**   Find $\sqrt{5 - 12i}$.

2 marks

ii   Hence, or otherwise, solve the equation $z^2 + 4z - 1 + 12i = 0$.

2 marks

b   Find $\displaystyle\int \frac{1}{\sqrt{x}\left(1 + \sqrt{x}\right)}\,dx$.

2 marks

Question 11 continues on page 126

Question 11 (continued)

c Evaluate $\int_{-1}^{2} \frac{dx}{x^2 + 2x + 10}$.   3 marks

d Consider the complex numbers $z = -\frac{1}{\sqrt{2}} + \frac{i}{\sqrt{2}}$ and $w = e^{-\frac{i\pi}{6}}$.

    i Express z and w in polar form.   2 marks

    ii Find w^7 in Cartesian form.   2 marks

Question 11 continues on page 127

Question 11 (continued)

iii Find $\arg\left(\dfrac{z}{w}\right)$ in exact form.   1 mark

iv Plot z and $\dfrac{1}{z}$ on an Argand diagram showing features.   1 mark

Question 12 (15 marks)

a Solve the differential equation $\sqrt{3x - x^2}\,\dfrac{dy}{dx} = 1 + \cos 2y$.   4 marks

b Use a counterexample to show that the following statement is false: 2 marks

$$\forall\, p, q \in \mathbb{R}, \; p - q > 0 \Rightarrow p^2 - q^2 > 0$$

Question 12 continues on page 128

Question 12 (continued)

c   Prove that $\forall\, x, y \in \mathbb{R}$ where $x, y > 0$, $x^3 + 2y^3 \geq 3xy^2$.   3 marks

d   A particle travels in a straight line according to the equation $\ddot{x} = t \sin t$, where x is its displacement   3 marks
     from O at time t. When $t = 0$, $x = 0$ and $v = V$.

     Find the equation for the velocity v in terms of t.

Question 12 continues on page 129

Question 12 (continued)

e The points $A(8, -1, 3)$ and $B(-2, 5, 7)$ are the endpoints of a diameter of a sphere. 3 marks

Find the equation of the sphere.

Question 13 (15 marks)

a Consider the vector equation $r = \begin{pmatrix} 1 \\ 0 \\ -2 \end{pmatrix} + \lambda \begin{pmatrix} 2 \\ -1 \\ 3 \end{pmatrix}, \lambda \in \mathbb{R}.$

i Find the vector equation of the line q which is parallel to r and passes through the point $C(5, 4, 2)$. 2 marks

ii Show that the vector $p = -i + j + k$ is perpendicular to r. 1 mark

iii Find the point of intersection between r and the line $l = \begin{pmatrix} 5 \\ -3 \\ -1 \end{pmatrix} + \lambda_2 p, \lambda_2 \in \mathbb{R}.$ 3 marks

Question 13 continues on page 130

Question 13 (continued)

b   *PQRS* is a quadrilateral with midpoints *K, L, M, N* of each side as shown.   3 marks

Prove that *MNKL* is a parallelogram.

Question 13 continues on page 131

Question 13 (continued)

c **i** By expanding $(\cos\theta + i\sin\theta)^3$, or otherwise, show that $\tan 3\theta = \dfrac{3\tan\theta - \tan^3\theta}{1 - 3\tan^2\theta}$.   2 marks

ii Hence, show that $\tan\left(\dfrac{\pi}{12}\right)$ is a solution to the equation $t^3 - 3t^2 - 3t + 1 = 0$.   4 marks

Find the other distinct solutions.

End of Question 13

Question 14 (15 marks)

a **i** Using the identity $z^9 - 1 = (z^3 - 1)(z^6 + z^3 + 1)$, or otherwise, find the 6 roots of $z^6 + z^3 + 1 = 0$ in polar form and plot them on an Argand diagram. 3 marks

ii Factorise $z^6 + z^3 + 1$ over the real plane. 3 marks

b Prove by contradiction that $x + \dfrac{1}{x} \geq 2$, $\forall x > 0$. 2 marks

Question 14 continues on page 133

Question 14 (continued)

c Consider the statement $P \Rightarrow Q$ for $n \in \mathbb{N}$:

'If n^2 is even, then n is even.'

   i Write the contrapositive statement to the above statement. 1 mark

   ii Use the contrapositive to prove that $P \Rightarrow Q$ is true. 2 marks

d **i** Show that the infinite series $1 - x^2 + x^4 - x^6 + x^8 - \cdots = \dfrac{1}{1 + x^2}$ for $|x| < 1$. 1 mark

Question 14 continues on page 134

Question 14 (continued)

ii   Consider the shaded area under the curve $y = \dfrac{1}{1+x^2}$ for $0 \leq x \leq 1$.   3 marks

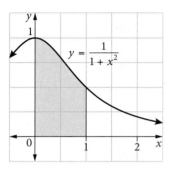

Use this area to show that $\dfrac{\pi^2}{4} \approx 1 - \dfrac{1}{3} + \dfrac{1}{5} - \dfrac{1}{7} + \dfrac{1}{9} - \ldots$

End of Question 14

Question 15 (15 marks)

a   **i**   Use a suitable substitution to prove that   2 marks

$$\int_0^a f(x)\,dx = \int_0^a f(a-x)\,dx.$$

ii   Hence, or otherwise, prove that   4 marks

$$\int_0^\pi \frac{x\sin x}{1+\cos^2 x}\,dx = \frac{\pi^2}{4}.$$

Question 15 continues on page 136

Question 15 (continued)

b   **i**   Show that $(k + 1)^3 - k^3 = 3k^2 + 3k + 1$.   1 mark

   ii   Find an expression for the sum $1 + 2 + 3 + \cdots + n$.   2 marks

   iii   By substituting $k = 1, 2, 3, \ldots, n$ into the formula from part **i** and adding, show that   3 marks

$$1^2 + 2^2 + 3^2 + \cdots + n^2 = \frac{n(n + 1)(2n + 1)}{6}.$$

Question 15 continues on page 137

Question 15 (continued)

iv   By expanding $(k + 1)^4 - k^4$ and using a similar method to that used in part **iii**, show that   3 marks

$$\sum_{k=1}^{n} k^3 = \frac{n^2(n+1)^2}{4}.$$

End of Question 15

Question 16 (15 marks)

a   A bullet of mass m kg is fired horizontally from $(0,0)$ at an initial speed of u m/s towards a target R metres to the right in the horizontal direction and H metres below in the vertical direction. The bullet undergoes acceleration due to gravity of g m/s^2 and air resistance of magnitude mv, where v is the velocity in m/s at time t seconds. You can assume that $\dot{\underset{\sim}{v}} = -\dot{x}\underset{\sim}{i} - (\dot{y} + g)\underset{\sim}{j}$ at time t seconds. (Do NOT prove this).

    i   Show that $\underset{\sim}{v} = ue^{-t}\underset{\sim}{i} + (ge^{-t} - g)\underset{\sim}{j}$ satisfies the equation for $\dot{\underset{\sim}{v}}$.   3 marks

    ii   Derive equations for the horizontal and vertical displacements x and y metres, respectively.   3 marks

Question 16 continues on page 139

Question 16 (continued)

    iii Show that if the bullet reaches the target, then $u = \dfrac{Rg}{gt - H}$.   2 marks

b   **i**   ©NESA   2017 HSC EXAM, QUESTION 16(a)

Let $\alpha = \cos\theta + i\sin\theta$, where $0 < \theta < 2\pi$.   1 mark

Show that $\alpha^k + \alpha^{-k} = 2\cos k\theta$ for any integer k.

    ii Furthermore, $C = \alpha^{-n} + \cdots + \alpha^{-1} + 1 + \alpha + \cdots + \alpha^n$, where n is a positive integer.   3 marks

By summing the series, prove that $C = \dfrac{\alpha^n + \alpha^{-n} - (\alpha^{n+1} + \alpha^{-(n+1)})}{(1 - \alpha)(1 - \bar{\alpha})}$.

Question 16 continues on page 140

Question 16 (continued)

   iii Deduce, from parts **i** and **ii**, that   2 marks

$$1 + 2(\cos\theta + \cos 2\theta + \cos 3\theta + \cdots + \cos n\theta) = \frac{\cos n\theta - \cos(n+1)\theta}{1 - \cos\theta}.$$

   iv Show that $\cos\left(\dfrac{\pi}{n}\right) + \cos\left(\dfrac{2\pi}{n}\right) + \cos\left(\dfrac{3\pi}{n}\right) + \cdots + \cos\left(\dfrac{n\pi}{n}\right)$ is independent of n.   1 mark

END OF PAPER

SECTION II EXTRA WORKING SPACE

WORKED SOLUTIONS

Section I (1 mark each)

Question 1

C $\int \dfrac{4-x}{(x-1)(x+2)}\,dx = \ln|x-1| - 2\ln|x+2| + c$

$$\frac{d}{dx}\Big(\ln|x-1| - 2\ln|x+2|\Big) = \frac{1}{x-1} - \frac{2}{x+2}$$

$$= \frac{1(x+2) - 2(x-1)}{(x-1)(x+2)}$$

$$= \frac{-x+4}{(x-1)(x+2)}$$

Note that this question can be done by differentiating the options, as well as decomposing the original question into partial fractions, then integrating.

Question 2

C Solving for λ using the point $(-1, 0, 0)$ gives a consistent value for λ:

$$1 - 2\lambda = -1 \rightarrow \lambda = 1$$

$$-1 + \lambda = 0 \rightarrow \lambda = 1$$

$$1 - \lambda = 0 \rightarrow \lambda = 1$$

All 3 equations must be satisfied.

Question 3

D If $z^5 + 1 = 0$, then one solution is $z = -1$. The only option is the one below.

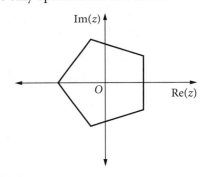

This question can be answered by a process of elimination of the options.

Question 4

B Since P is oscillating about O and is stationary at $x = 5$, then the amplitude is 5. If it travels from $x = 5$ to O after 3 seconds, a quarter of the period, then period is $4 \times 3 = 12$ seconds.

$$\text{Period} = \frac{2\pi}{n} = 12$$

$$n = \frac{2\pi}{12}$$

$$= \frac{\pi}{6}$$

Let $x = A \sin(nt + \alpha) = 5\sin\left(\dfrac{\pi t}{6} + \alpha\right)$.

$$\dot{x} = \frac{5\pi}{6}\cos\left(\frac{\pi t}{6} + \alpha\right)$$

The range of $\dot{x}$ is $\left[-\dfrac{5\pi}{6}, \dfrac{5\pi}{6}\right]$ so the maximum speed $|\dot{x}|$ is $\dfrac{5\pi}{6}$.

Having an understanding of the physical movement under simple harmonic motion, rather than relying on equations, can save a lot of time and effort.

Question 5

D The negation is: 'Not all composite numbers have more than two factors', which is the same as:

'At least one composite number does not have more than two factors.'

To negate a statement, put a 'not' in it. The negation of 'All are' is 'at least one is not'.

Question 6

D The coordinates of the new point are $(5, 2, 7)$ since only the y-value will change as the others are invariant.

Drawing a rough sketch can help here.

Question 7

B Use a sign diagram with the critical values to

solve $\dfrac{x+1}{(x-1)(x+3)} \geq 0$:

$$\begin{array}{ccccc} - & + & - & + \\ \hline & \circ & \bullet & \circ & \\ -3 & -1 & 1 & & x \end{array}$$

$-3 < x \leq -1$ or $x > 1$

> Watch the discontinuities in the denominator.

Question 8

C $\dfrac{d}{dx}\left(\dfrac{1}{2}v^2\right) = 4 - x$

$$\dfrac{1}{2}v^2 = 4x - \dfrac{1}{2}x^2 + k$$

$$v^2 = 8x - x^2 + c$$

$$v = \pm\sqrt{8x - x^2 + c}$$

Therefore, $\dot{x} = \sqrt{8x - x^2 + c}$ is the only possible option.

> To integrate $\ddot{x} = 4 - x$, we need to use the
> identity $\ddot{x} = \dfrac{d}{dx}\left(\dfrac{1}{2}v^2\right)$ since the equation given
> is in terms of x not t. Sometimes, it can be
> determined to take the positive or negative case
> if the initial conditions are given but they are not
> in this question.

Question 9

B It is important to recognise that

$$\int_0^2 \sqrt{x(2-x)}\,dx = \int_0^2 \sqrt{2x - x^2}\,dx$$

$$= \int_0^2 \sqrt{-x^2 + 2x - 1 + 1}\,dx$$

$$= \int_0^2 \sqrt{-(x^2 - 2x + 1) + 1}\,dx$$

$$= \int_0^2 \sqrt{-(x-1)^2 + 1}\,dx$$

$$= \int_0^2 \sqrt{1 - (x-1)^2}\,dx$$

which represents the area under a semicircle with centre $(1, 0)$ and radius 1.

$$\int_0^2 \sqrt{x(2-x)}\,dx = \dfrac{1}{2}\pi(1^2)$$

$$= \dfrac{\pi}{2}$$

Alternative solution for evaluating

$\int_0^2 \sqrt{1 - (x-1)^2}\,dx$:

Let $x - 1 = \sin u$, then $dx = \cos u\,du$.

When $x = 0$, $u = -\dfrac{\pi}{2}$, and when $x = 2$, $u = \dfrac{\pi}{2}$.

$$\int_0^2 \sqrt{1 - (x-1)^2}\,dx$$

$$= \int_{-\frac{\pi}{2}}^{\frac{\pi}{2}} \sqrt{1 - \sin^2 u} \times \cos u\,du$$

$$= \int_{-\frac{\pi}{2}}^{\frac{\pi}{2}} \cos^2 u\,du$$

$$= \dfrac{1}{2}\int_{-\frac{\pi}{2}}^{\frac{\pi}{2}} 1 + \cos 2u\,du$$

$$= \dfrac{1}{2}\left[u + \dfrac{1}{2}\sin 2u\right]_{-\frac{\pi}{2}}^{\frac{\pi}{2}}$$

$$= \dfrac{\pi}{2}$$

> This is a difficult question. Always look for
> semicircles or quadrants. The answer can
> be seen through a sketch.

Question 10

A Consider the parallelogram formed by O, $z_1, z_1 + z_2, z_2$.

If $\arg\left(\dfrac{z_1 + z_2}{z_2 - z_1}\right) = \dfrac{\pi}{2}$ then the angle between

the two diagonals using the parallelogram rule must be 90°. This means that the parallelogram must be a rhombus. All sides are equal. Therefore $|z_1| = |z_2|$.

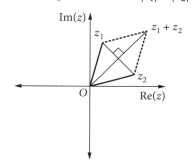

> Note the other options are not correct since
> the diagonals are not necessarily equal and
> rotation must be in the anticlockwise direction
> for positive arguments.

Section II ($\checkmark$ = 1 mark)

Question 11 (15 marks)

a i $\sqrt{5 - 12i} = a + ib$, where $a, b \in \mathbb{R}$

$5 - 12i = a^2 - b^2 + 2abi$

Equating real and imaginary parts:

$5 = a^2 - b^2, -6 = ab$ $\checkmark$

By inspection, we see that
$a = 3, b = -2$ or $a = -3, b = 2$.

So $\sqrt{5 - 12i} = \pm(3 - 2i)$. $\checkmark$

> It is important to state that a and b are real to eliminate other solutions.

ii $z^2 + 4z - 1 + 12i = 0$

$z = \dfrac{-b \pm \sqrt{b^2 - 4ac}}{2a}$

$= \dfrac{-4 \pm \sqrt{4^2 - 4(1)(-1 + 12i)}}{2(1)}$ $\checkmark$

$= \dfrac{-4 \pm \sqrt{20 - 48i}}{2}$

$= \dfrac{-4 \pm 2\sqrt{5 - 12i}}{2}$

$= \dfrac{-4 \pm 2(3 - 2i)}{2}$

$= \dfrac{2 - 4i}{2}$ OR $\dfrac{-10 + 4i}{2}$

$= 1 - 2i$ OR $-5 + 2i$ $\checkmark$

> The solution to part **i** is needed to find the solutions in part **ii**. Sometimes a question like this is NOT scaffolded into 2 parts with clues so you have to work it all out yourself. See 2020 HSC exam, Question 11(e).

b Check derivative of $y = 1 + \sqrt{x}$.

$\dfrac{dy}{dx} = \dfrac{1}{2}x^{-\frac{1}{2}} = \dfrac{1}{2\sqrt{x}}$

$\displaystyle\int \dfrac{1}{\sqrt{x}(1 + \sqrt{x})} dx = 2\int \dfrac{1}{2\sqrt{x}} \dfrac{1}{(1 + \sqrt{x})} dx$ $\checkmark$

$= 2\ln(1 + \sqrt{x}) + c$ $\checkmark$

> Always look out for integrals of the form $\int \dfrac{f'(x)}{f(x)} dx$, which become logarithmic functions.
> If in doubt, try the substitution $u = 1 + \sqrt{x}$.
> It's not immediately obvious that $\dfrac{1}{2\sqrt{x}}$ is part of the derivative of $\sqrt{x}$ but it's a result worth remembering.

c $\displaystyle\int_{-1}^{2} \dfrac{dx}{x^2 + 2x + 10} = \int_{-1}^{2} \dfrac{dx}{(x + 1)^2 + 9}$ $\checkmark$

$= \dfrac{1}{3}\left[\tan^{-1}\left(\dfrac{x + 1}{3}\right)\right]_{-1}^{2}$ $\checkmark$

$= \dfrac{1}{3}\left[\tan^{-1}\left(\dfrac{2 + 1}{3}\right) - \tan^{-1}\left(\dfrac{-1 + 1}{3}\right)\right]$

$= \dfrac{1}{3}\left(\tan^{-1}1 - \tan^{-1}0\right)$

$= \dfrac{1}{3}\left(\dfrac{\pi}{4} - 0\right)$

$= \dfrac{\pi}{12}$ $\checkmark$

> The denominator does not factorise so completing the square is the only option. Look at the HSC exam reference sheet if you are unsure of the integrals that give inverse trigonometric functions.

d i $z = -\dfrac{1}{\sqrt{2}} + \dfrac{i}{\sqrt{2}}$

$|z| = \sqrt{\left(-\dfrac{1}{\sqrt{2}}\right)^2 + \left(\dfrac{1}{\sqrt{2}}\right)^2}$

$= \sqrt{\dfrac{1}{2} + \dfrac{1}{2}}$

$= \sqrt{1}$

$= 1$

$\tan\theta = \dfrac{\dfrac{1}{\sqrt{2}}}{-\dfrac{1}{\sqrt{2}}} = -1$

$\theta = \dfrac{3\pi}{4}$ since z is in the 2nd quadrant.

$z = \cos\dfrac{3\pi}{4} + i\sin\dfrac{3\pi}{4}$ $\checkmark$

$w = e^{-\frac{i\pi}{6}}$

$= \cos\left(-\dfrac{\pi}{6}\right) + i\sin\left(-\dfrac{\pi}{6}\right)$ $\checkmark$

> Polar form is modulus-argument form $r(\cos\theta + i\sin\theta)$. Note that the signs determine the quadrant.

ii $w^7 = \left[\cos\left(-\frac{\pi}{6}\right) + i\sin\left(-\frac{\pi}{6}\right)\right]^7$

$= \cos\left(-\frac{7\pi}{6}\right) + i\sin\left(-\frac{7\pi}{6}\right)$ ✔

$= \cos\frac{5\pi}{6} + i\sin\frac{5\pi}{6}$

$= -\frac{\sqrt{3}}{2} + \frac{i}{2}$ ✔

Note how polar form simplifies multiplying and dividing complex numbers.

iii $\arg\left(\frac{z}{w}\right) = \arg z - \arg w$

$= \frac{3\pi}{4} - \left(-\frac{\pi}{6}\right)$

$= \frac{11\pi}{12}$ ✔

Notice how the argument laws are similar to the logarithm laws.

iv $\frac{1}{z} = z^{-1} = \cos\left(-\frac{3\pi}{4}\right) + i\sin\left(-\frac{3\pi}{4}\right)$.

Note modulus of both z and $\frac{1}{z}$ is 1.

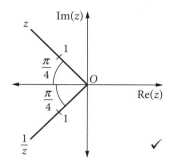

Diagram must show equal moduli and angles. The reciprocal reflects the argument in the x-axis.

Question 12 (15 marks)

a $\sqrt{3x - x^2}\,\frac{dy}{dx} = 1 + \cos 2y$

$\frac{dy}{1 + \cos 2y} = \frac{dx}{\sqrt{3x - x^2}}$

$\frac{dy}{2\cos^2 y} = \frac{dx}{\sqrt{\left(\frac{3}{2}\right)^2 - \left(x^2 - 3x + \left(\frac{3}{2}\right)^2\right)}}$,   (note: $\cos 2y = 2\cos^2 y - 1$) ✔

$\int \frac{1}{2}\sec^2 y\,dy = \int \frac{dx}{\sqrt{\left(\frac{3}{2}\right)^2 - \left(x - \frac{3}{2}\right)^2}}$ ✔

$\frac{1}{2}\tan y = \sin^{-1}\left(\frac{x - \frac{3}{2}}{\frac{3}{2}}\right) + k$ ✔

$y = \tan^{-1}\left[2\sin^{-1}\left(\frac{2}{3}x - 1\right) + c\right]$ ✔

Separating the variables on the LHS and RHS makes the techniques to employ easier to recognise. Complete the square in the denominator to create an integral that gives an inverse sine function.

b $\forall p, q \in \mathbb{R}, p - q > 0 \Rightarrow p^2 - q^2 > 0$.

Let $p = 1, q = -2$

Then $p - q = 1 - (-2) = 3 > 0$, which makes the 'if' part of the statement true. ✔

$p^2 - q^2 = 1^2 - (-2)^2 = -3 < 0$, which makes the statement false.

So the statement is not always true. ✔

Only one example is required to prove an incorrect statement false.

c   RTP that $\forall x, y \in \mathbb{R}$, where $x, y > 0$, $x^3 + 2y^3 \geq 3xy^2$.

Proof:

Consider the difference:

$$\begin{aligned}
x^3 + 2y^3 - 3xy^2 &= x^3 - xy^2 + 2y^3 - 2xy^2 \\
&= x(x^2 - y^2) + 2y^2(y - x) \checkmark \\
&= x(x + y)(x - y) + 2y^2(y - x) \\
&= (x - y)[x(x + y) - 2y^2] \\
&= (x - y)(x^2 + xy - 2y^2) \checkmark \\
&= (x - y)(x + 2y)(x - y) \\
&= (x - y)^2(x + 2y) \\
&\geq 0 \qquad \text{because} \\
&\qquad (x - y)^2 \geq 0, (x + 2y) > 0.
\end{aligned}$$

So $x^3 + 2y^3 \geq 3xy^2$. $\checkmark$

> The 'consider the difference' proof for inequalities is based on the definition that $a > b$ iff $a - b > 0$.

d $\ddot{x} = t \sin t$

$\dot{x} = \int t \sin t \, dt$

Using integration by parts:

$u = t \qquad\qquad v' = \sin t$

$u' = 1 \qquad\qquad v = -\cos t$ $\checkmark$

$$\begin{aligned}
\dot{x} &= -t \cos t - \int -\cos t \, d \\
&= -t \cos t + \int \cos t \, dt \\
&= -t \cos t + \sin t + c \checkmark
\end{aligned}$$

When $t = 0$, $v = V$, so substituting:

$V = -0 \cos 0 + \sin 0 + c$

$V = c$

$\therefore v = -t \cos t + \sin t + V$ $\checkmark$

> It is a good idea to state the formula or the technique you have chosen so that the examiner can follow your solution (in case you make an error).

e   Midpoint of $A(8, -1, 3)$ and $B(-2, 5, 7)$ is the centre of the sphere.

$$C(x, y, z) = \left(\frac{8 + -2}{2}, \frac{-1 + 5}{2}, \frac{3 + 7}{2}\right) = (3, 2, 5) \checkmark$$

$$\begin{aligned}
\text{Diameter} &= \sqrt{(8 - -2)^2 + (-1 - 5)^2 + (3 - 7)^2} \\
&= \sqrt{152} \\
&= 2\sqrt{38}
\end{aligned}$$

Radius $= \sqrt{38}$ $\checkmark$

Equation of the sphere:

$$(x - 3)^2 + (y - 2)^2 + (z - 5)^2 = 38 \checkmark$$

> The same technique used for 2D coordinate geometry is extended to 3D.

Question 13 (15 marks)

a   **i**   The line parallel to $\underset{\sim}{r} = \begin{pmatrix} 1 \\ 0 \\ -2 \end{pmatrix} + \lambda \begin{pmatrix} 2 \\ -1 \\ 3 \end{pmatrix}$

and through $A(5, 4, 2)$ has the same direction vector. $\checkmark$

$$\underset{\sim}{q} = \begin{pmatrix} 5 \\ 4 \\ 2 \end{pmatrix} + \lambda \begin{pmatrix} 2 \\ -1 \\ 3 \end{pmatrix}, \text{ where } \lambda \in \mathbb{R}. \checkmark$$

> Score 1 mark for using the same direction vector or equivalent merit. The direction vector in 3 dimensions is similar to the gradient in 2 dimensions.

   ii   Show $\underset{\sim}{p} = -\underset{\sim}{i} + \underset{\sim}{j} + \underset{\sim}{k}$ and $\begin{pmatrix} 2 \\ -1 \\ 3 \end{pmatrix}$ are

perpendicular, that is, dot product $= 0$.

$$\begin{aligned}
\begin{pmatrix} -1 \\ 1 \\ 1 \end{pmatrix} \cdot \begin{pmatrix} 2 \\ -1 \\ 3 \end{pmatrix} &= -1 \times 2 + 1 \times (-1) + 1 \times 3 \\
&= -2 - 1 + 3 \\
&= 0 \checkmark
\end{aligned}$$

iii $\underset{\sim}{l} = \begin{pmatrix} 5 \\ -3 \\ -1 \end{pmatrix} + \lambda_2 \underset{\sim}{p}, \ \lambda_2 \in \mathbb{R}$

where $\underset{\sim}{p} = -\underset{\sim}{i} + \underset{\sim}{j} + \underset{\sim}{k}$.

That is, $\underset{\sim}{l} = \begin{pmatrix} 5 \\ -3 \\ -1 \end{pmatrix} + \lambda_2 \begin{pmatrix} -1 \\ 1 \\ 1 \end{pmatrix}$.

Let the point (a, b, c) be the point of intersection between $\underset{\sim}{l}$ and $\underset{\sim}{r}$, so it satisfies both.

$\begin{pmatrix} a \\ b \\ c \end{pmatrix} = \begin{pmatrix} 1 \\ 0 \\ -2 \end{pmatrix} + \lambda \begin{pmatrix} 2 \\ -1 \\ 3 \end{pmatrix}$ and

$\begin{pmatrix} a \\ b \\ c \end{pmatrix} = \begin{pmatrix} 5 \\ -3 \\ -1 \end{pmatrix} + \lambda_2 \begin{pmatrix} -1 \\ 1 \\ 1 \end{pmatrix}$,

so equating

$\begin{pmatrix} 1 \\ 0 \\ -2 \end{pmatrix} + \lambda \begin{pmatrix} 2 \\ -1 \\ 3 \end{pmatrix} = \begin{pmatrix} 5 \\ -3 \\ -1 \end{pmatrix} + \lambda_2 \begin{pmatrix} -1 \\ 1 \\ 1 \end{pmatrix}$. ✔

$1 + 2\lambda = 5 - \lambda_2$ [1]

$0 - \lambda = -3 + \lambda_2 \Rightarrow \lambda_2 = 3 - \lambda$

$-2 + 3\lambda = -1 + \lambda_2$ [2]

Check $\lambda_2 = 3 - \lambda$ into [1] and [2]:

$\lambda_2 = 3 - \lambda$

$1 + 2\lambda = 5 - (3 - \lambda) \ [1] \Rightarrow \lambda = 1$

$-2 + 3\lambda = -1 + 3 - \lambda \ [2] \Rightarrow \lambda = 1$

So $\lambda = 1$ and $\lambda_2 = 3 - \lambda = 3 - 1 = 2$. ✔

The point of intersection is:

$\begin{pmatrix} a \\ b \\ c \end{pmatrix} = \begin{pmatrix} 1 \\ 0 \\ -2 \end{pmatrix} + 1 \begin{pmatrix} 2 \\ -1 \\ 3 \end{pmatrix} = \begin{pmatrix} 3 \\ -1 \\ 1 \end{pmatrix}$ ✔

The solution must satisfy all 3 equations.

b Let the vectors be labelled as shown.
M, N, K, L are midpoints.

RTP: $MNKL$ is a parallelogram.

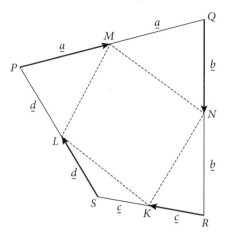

Proof:

$\overline{LM} = \underset{\sim}{d} + \underset{\sim}{a}$ and $\overline{KN} = -\underset{\sim}{c} + -\underset{\sim}{b}$ ✔

Note that $\overline{PQ} + \overline{QR} + \overline{RS} + \overline{SP} = 2\underset{\sim}{a} + 2\underset{\sim}{b} + 2\underset{\sim}{c} + 2\underset{\sim}{d}$.

But $\overline{PQ} + \overline{QR} + \overline{RS} + \overline{QP} = \overline{PP} = 0$.

$\therefore 2\underset{\sim}{a} + 2\underset{\sim}{b} + 2\underset{\sim}{c} + 2\underset{\sim}{d} = 0$

$\underset{\sim}{a} + \underset{\sim}{d} = -\underset{\sim}{b} - \underset{\sim}{c}$ ✔

So $\overline{LM} = \overline{KN}$.

Therefore, $MNKL$ is a parallelogram (one pair of opposite sides parallel and equal). QED. ✔

Adding vectors head to tail is an important skill, noting the direction. Also, know your properties of quadrilaterals from Years 7–10 because vectors are applied to geometrical proofs in Mathematics Extension 1 and 2.

c **i** Using De Moivre's theorem $(\cos\theta + i\sin\theta)^3 = \cos 3\theta + i\sin 3\theta$.

Using the binomial theorem, $(\cos\theta + i\sin\theta)^3 = \cos^3\theta + 3i\cos^2\theta\sin\theta - 3\cos\theta\sin^2\theta - i\sin^3\theta$.

Equating real and imaginary parts:

$\cos 3\theta = \cos^3\theta - 3\cos\theta\sin^2\theta$
$\sin 3\theta = 3\cos^2\theta\sin\theta - \sin^3\theta$ ✓

So dividing:

$$\tan 3\theta = \frac{\sin 3\theta}{\cos 3\theta}$$

$$= \frac{3\cos^2\theta\sin\theta - \sin^3\theta}{\cos^3\theta - 3\cos\theta\sin^2\theta} \div \frac{\cos^3\theta}{\cos^3\theta}$$

$$= \frac{\dfrac{3\sin\theta}{\cos\theta} - \dfrac{\sin^3\theta}{\cos^3\theta}}{1 - \dfrac{3\sin^2\theta}{\cos^2\theta}}$$

$\therefore \tan 3\theta = \dfrac{3\tan\theta - \tan^3\theta}{1 - 3\tan^2\theta}$, as required. ✓

> This is a very typical exam question using the expansion of $\cos\theta + i\sin\theta$ in 2 different ways to prove various identities. This formula can also be derived using the $\tan(A + B)$ results.

ii Rewrite $t^3 - 3t^2 - 3t + 1 = 0$ into the $\tan 3\theta$ form:

$-3t^2 + 1 = 3t - t^3$

$1 = \dfrac{3t - t^3}{1 - 3t^2}$ ✓

Substitute $t = \tan\dfrac{\pi}{12}$ into the RHS, then use our result in part **i**:

$$\text{RHS} = \frac{3\tan\dfrac{\pi}{12} - \tan^3\dfrac{\pi}{12}}{1 - 3\tan^2\dfrac{\pi}{12}}$$

$$= \tan 3\left(\frac{\pi}{12}\right) \quad \text{(using the identity in part } \mathbf{i}\text{)}$$

$$= \tan\frac{\pi}{4}$$

$$= 1$$

$$= \text{LHS.}$$

So $\tan\dfrac{\pi}{12}$ is a solution to the equation $t^3 - 3t^2 - 3t + 1 = 0$. ✓

For the other solutions, let $t = \tan\theta$ and solve: $\dfrac{3\tan\theta - \tan^3\theta}{1 - 3\tan^2\theta} = 1$

$\tan 3\theta = 1$

$3\theta = \dfrac{\pi}{4}, \dfrac{5\pi}{4}, \dfrac{\pi}{4} \pm 2\pi, \dfrac{5\pi}{4} \pm 2\pi, \cdots$

$\therefore \theta = \dfrac{\pi}{12}, \dfrac{5\pi}{12}, -\dfrac{7\pi}{12}, -\dfrac{3\pi}{12} = -\dfrac{\pi}{4}, \cdots$ ✓

So $\tan\dfrac{\pi}{12}$ is a root. Other distinct roots are $\tan\dfrac{5\pi}{12}$, $\tan\left(-\dfrac{\pi}{4}\right) = -1$ since the others are repeats of these 3. ✓

> As this is a cubic equation, it has at most 3 roots. This question relies on identifying the relationship between a polynomial and a trigonometric equation to solve. It is necessary to list a number of solutions in order to identify a unique set.

Question 14 (15 marks)

a i The roots of $z^9 - 1 = (z^3 - 1)(z^6 + z^3 + 1) = 0$ are equally spaced around the Argand diagram, with the 3 cube roots of unity, z_3, z_6 and 1, being the roots of $(z^3 - 1) = 0$.
The other 6 roots, z_1, z_2, z_4, z_5, z_7 and z_8, must be roots of $z^6 + z^3 + 1 = 0$.

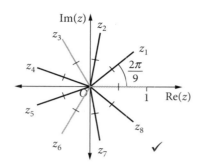

$z_1 = \cos\dfrac{2\pi}{9} + i\sin\dfrac{2\pi}{9}, z_2 = \cos\dfrac{4\pi}{9} + i\sin\dfrac{4\pi}{9}, z_3 = \cos\dfrac{6\pi}{9} + i\sin\dfrac{6\pi}{9},$

$z_4 = \cos\dfrac{8\pi}{9} + i\sin\dfrac{8\pi}{9}, z_5 = \cos\left(-\dfrac{8\pi}{9}\right) + i\sin\left(-\dfrac{8\pi}{9}\right), z_6 = \cos\left(-\dfrac{6\pi}{9}\right) + i\sin\left(-\dfrac{6\pi}{9}\right),$

$z_7 = \cos\left(-\dfrac{4\pi}{9}\right) + i\sin\left(-\dfrac{4\pi}{9}\right), z_8 = \cos\left(-\dfrac{2\pi}{9}\right) + i\sin\left(-\dfrac{2\pi}{9}\right), z_9 = 1$ ✔✔

It is useful to learn the factorisation of a sum or difference of 2 cubes and also of $a^n - b^n$.
Drawing a diagram is always encouraged to relate the algebra to the geometric context.

ii $z^6 + z^3 + 1 = (z - z_1)(z - z_2)(z - z_4)(z - z_5)(z - z_7)(z - z_8)$

$= (z - z_1)(z - z_2)(z - z_4)(z - \overline{z_4})(z - \overline{z_2})(z - \overline{z_1})$

$= [z^2 - (z_1 + \overline{z_1})z + 1][z^2 - (z_2 + \overline{z_2})z + 1][z^2 - (z_4 + \overline{z_4})z + 1]$ ✔   (by grouping conjugates)

Now, note that

$z_1 + \overline{z_1} = \left(\cos\dfrac{2\pi}{9} + i\sin\dfrac{2\pi}{9}\right) + \left[\cos\left(-\dfrac{2\pi}{9}\right) + i\sin\left(-\dfrac{2\pi}{9}\right)\right]$

$= \left(\cos\dfrac{2\pi}{9} + i\sin\dfrac{2\pi}{9}\right) + \left(\cos\dfrac{2\pi}{9} - i\sin\dfrac{2\pi}{9}\right)$

$= 2\cos\dfrac{2\pi}{9}$ etc. ✔

So simplifying,

$z^6 + z^3 + 1 = \left[z^2 - \left(2\cos\dfrac{2\pi}{9}\right)z + 1\right]\left[z^2 - \left(2\cos\dfrac{4\pi}{9}\right)z + 1\right]\left[z^2 - \left(2\cos\dfrac{8\pi}{9}\right)z + 1\right].$ ✔

Knowing the identity $(x - a)(x - b) = x^2 - (a + b)x + ab$ saves a lot of tedious algebra.

b $\forall x > 0, x + \dfrac{1}{x} \geq 2$

Proof by contradiction:

Assume that $\exists x > 0, x + \dfrac{1}{x} < 2$.

Therefore,

$$x + \frac{1}{x} - 2 < 0$$

$$\frac{x^2 + 1 - 2x}{x} < 0$$

$$\frac{(x-1)^2}{x} < 0. \checkmark$$

Now $(x-1)^2 \geq 0$ and $x > 0$, so $\dfrac{(x-1)^2}{x} > 0$.

Contradiction.

Therefore, $\forall x > 0, x + \dfrac{1}{x} \geq 2. \checkmark$

> This inequality is well known and can also be proved by considering the difference.

c **i**   The contrapositive is $\neg Q \Rightarrow \neg P$:
'If n is not even, then n^2 is not even.' $\checkmark$

ii If n is not even then n is odd since n is a positive integer by definition.

If n is odd then n^2 is also odd. This is true, as the product of 2 odd numbers is odd.

Therefore, the contrapositive $\neg Q \Rightarrow \neg P$ is true. $\checkmark$

Since the contrapositive is true then the original implication $P \Rightarrow Q$ is also true since they are equivalent. $\checkmark$

d **i** $|x| < 1$, $1 - x^2 + x^4 - x^6 + x^8 - \ldots$ is an infinite geometric series with a limiting sum, where $a = 1$ and $r = \dfrac{-x^2}{1} = \dfrac{x^4}{-x^2} = -x^2$.

$$\therefore S_\infty = \frac{a}{1-r} = \frac{1}{1-(-x^2)} = \frac{1}{1+x^2}$$

$$1 - x^2 + x^4 - x^6 + x^8 - \ldots = \frac{1}{1+x^2} \checkmark$$

ii Consider the area under the curve:

$$\int_0^1 \frac{1}{1+x^2} \, dx = \left[\tan^{-1} x \right]_0^1$$

$$= \tan^{-1} 1 - \tan^{-1} 0$$

$$= \frac{\pi}{4} - 0$$

$$= \frac{\pi}{4} \checkmark$$

But also from part **i**:

$$\int_0^1 \frac{1}{1+x^2} \, dx$$

$$= \int_0^1 1 - x^2 + x^4 - x^6 + x^8 - \ldots \, dx$$

$$= \left[x - \frac{x^3}{3} + \frac{x^5}{5} - \frac{x^7}{7} + \frac{x^9}{9} - \ldots \right]_0^1$$

$$= 1 - \frac{1^3}{3} + \frac{1^5}{5} - \frac{1^7}{7} + \frac{1^9}{9} - \cdots - (0 - 0 + 0 - \ldots)$$

$$= 1 - \frac{1}{3} + \frac{1}{5} - \frac{1}{7} + \frac{1}{9} - \ldots \checkmark$$

$$\therefore \frac{\pi}{4} \approx 1 - \frac{1}{3} + \frac{1}{5} - \frac{1}{7} + \frac{1}{9} - \ldots \checkmark$$

> Note that this part of the question is related to part **i**. It would be unusual if the parts of a question were not related. This type of question is quite typical of later questions in the Extension 2 exam, where a sum of areas of rectangles under a curve is related to the integral.

WORKED SOLUTIONS

Question 15 (15 marks)

a i Let $u = a - x$,

then $du = -dx$.

If $x = 0$, $u = a$.
If $x = a$, $u = 0$. ✔

$$\therefore \int_0^a f(a - x)\,dx = \int_a^0 -f(u)\,du$$

$$= \int_0^a f(u)\,du$$

$$= \int_0^a f(x)\,dx$$

So $\int_0^a f(x)\,dx = \int_0^a f(a - x)\,dx$. ✔

$\int_0^a f(u)\,du$ can be rewritten as $\int_0^a f(x)\,dx$ because u and x are just variables of integration. It is worth knowing this formula as it often arises in Maths Extension 2 integration problems.

ii If $\int_0^\pi f(x)\,dx = \int_0^\pi f(\pi - x)\,dx$ then:

$$\int_0^\pi \frac{x \sin x}{1 + \cos^2 x}\,dx = \int_0^\pi \frac{(\pi - x)\sin(\pi - x)}{1 + \cos^2(\pi - x)}\,dx$$

$$= \int_0^\pi \frac{\pi \sin(\pi - x) - x\sin(\pi - x)}{1 + \cos^2(\pi - x)}\,dx \quad ✔$$

$$= \int_0^\pi \frac{\pi \sin x - x\sin x}{1 + (-\cos x)^2}\,dx \text{ since } \sin(\pi - \theta) = \sin\theta, \cos(\pi - \theta) = -\cos\theta$$

$$= \int_0^\pi \frac{\pi \sin x - x\sin x}{1 + \cos^2 x}\,dx$$

$$= \int_0^\pi \frac{\pi \sin x}{1 + \cos^2 x} - \frac{x\sin x}{1 + \cos^2 x}\,dx \quad ✔$$

Note that the 2nd term on the RHS is the same as the LHS.

$$\therefore 2\int_0^\pi \frac{x\sin x}{1 + \cos^2 x}\,dx = \int_0^\pi \frac{\pi \sin x}{1 + \cos^2 x}\,dx$$

$$\therefore \int_0^\pi \frac{x\sin x}{1 + \cos^2 x}\,dx = \frac{1}{2}\int_0^\pi \frac{\pi \sin x}{1 + \cos^2 x}\,dx \quad ✔$$

$$= -\frac{\pi}{2}\int_0^\pi \frac{(-\sin x)}{1 + (\cos x)^2}\,dx$$

$$= -\frac{\pi}{2}\Big[\tan^{-1}(\cos x)\Big]_0^\pi$$

$$= -\frac{\pi}{2}\Big[\tan^{-1}(\cos \pi) - \tan^{-1}(\cos 0)\Big]$$

$$= -\frac{\pi}{2}\Big[\tan^{-1}(-1) - \tan^{-1}(1)\Big]$$

$$= -\frac{\pi}{2}\left(-\frac{\pi}{4} - \frac{\pi}{4}\right)$$

$$= -\frac{\pi}{2}\left(-\frac{\pi}{2}\right)$$

$$= \frac{\pi^2}{4} \quad ✔$$

b **i** $(k + 1)^3 - k^3 = k^3 + 3k^2 + 3k + 1 - k^3$
$$= 3k^2 + 3k + 1 \checkmark$$

ii $1 + 2 + 3 + \cdots + n$ is an arithmetic series with $a = 1$, $d = 1$ and n terms. $\checkmark$

$$S_n = \frac{n}{2}(a + l)$$
$$= \frac{n}{2}(1 + n) \checkmark$$

iii Using the formula in part **i** with $k = 1, 2, 3, \ldots, n$ and adding them:

$$(2)^3 - 1^3 = 3(1)^2 + 3(1) + 1$$
$$(3)^3 - 2^3 = 3(2)^2 + 3(2) + 1$$
$$(4)^3 - 3^3 = 3(3)^2 + 3(3) + 1$$
$$\vdots$$
$$(n + 1)^3 - n^3 = 3(n)^2 + 3(n) + 1 \checkmark$$

$$\therefore (n + 1)^3 - 1^3 = 3(1^2 + 2^2 + 3^2 + \cdots + n^2) + 3(1 + 2 + 3 + \cdots + n) + n$$

$$= 3(1^2 + 2^2 + 3^2 + \cdots + n^2) + 3\left[\frac{n}{2}(1 + n)\right] + n \quad \text{(from part ii)}$$

$$(n + 1)^3 - 1 - n = 3(1^2 + 2^2 + 3^2 + \cdots + n^2) + \frac{3n(n + 1)}{2}$$

$$\frac{(n + 1)^3 - 1 - n}{3} = 1^2 + 2^2 + 3^2 + \cdots + n^2 + \frac{n(n + 1)}{2}$$

$$\frac{(n + 1)^3 - 1 - n}{3} - \frac{n(n + 1)}{2} = 1^2 + 2^2 + 3^2 + \cdots + n^2$$

$$1^2 + 2^2 + 3^2 + \cdots + n^2 = \frac{2\left[(n + 1)^3 - (n + 1)\right] - 3n(n + 1)}{6} \checkmark$$

$$= \frac{(n + 1)\left\{2\left[(n + 1)^2 - 1\right] - 3n\right\}}{6}$$

$$= \frac{(n + 1)\left[2\left(n^2 + 2n\right) - 3n\right]}{6}$$

$$= \frac{(n + 1)(2n^2 + n)}{6}$$

$$= \frac{n(n + 1)(2n + 1)}{6} \checkmark$$

This is a fairly complex and detailed proof. This identity can also be proved by induction, but the question does not ask for this. This technique of writing subsequent terms and adding LHS and RHS to derive a sophisticated result, such as the sum of squares, is often tested in the Maths Extension 2 exams.

iv $\displaystyle\sum_{k=1}^{n} k^3 = 1^3 + 2^3 + 3^3 + \cdots + (n-1)^3 + n^3.$

Proof:

By expanding we find:

$$(k+1)^4 - k^4 = k^4 + 4k^3 + 6k^2 + 4k^2 + 1 - k^4$$
$$= 4k^3 + 6k^2 + 4k + 1$$

Substituting values of k and adding all formulas:

$$(2)^4 - 1^4 = 4(1)^3 + 6(1)^2 + 4(1) + 1$$

$$(3)^4 - 2^4 = 4(2)^3 + 6(2)^2 + 4(2) + 1$$

$$(4)^4 - 3^4 = 4(3)^3 + 6(3)^2 + 4(3) + 1$$

$$\vdots$$

$$(n+1)^4 - n^4 = 4(n)^3 + 6(n)^2 + 4(n) + 1 \quad \checkmark$$

$$(n+1)^4 - 1^4 = 4(1^3 + 2^3 + 3^3 + \cdots + n^3) + 6(1^2 + 2^2 + 3^2 + \cdots + n^2) + 4(1 + 2 + 3 + \cdots + n) + n$$

$$\therefore (n+1)^4 - 1 = 4\left(\sum_{k=1}^{n} k^3\right) + 6\left[\frac{n(n+1)(2n+1)}{6}\right] + 4\left[\frac{n}{2}(1+n)\right] + n \quad \text{(from parts \textbf{ii} and \textbf{iii})}$$

$$(n+1)^4 - 1 - n = 4\left(\sum_{k=1}^{n} k^3\right) + n(n+1)(2n+1) + 2n(1+n)$$

$$4\left(\sum_{k=1}^{n} k^3\right) = (n+1)^4 - 1 - n - n(n+1)(2n+1) - 2n(1+n)$$

$$\sum_{k=1}^{n} k^3 = \frac{(n+1)^4 - (n+1) - n(n+1)(2n+1) - 2n(1+n)}{4} \quad \checkmark$$

$$= \frac{(n+1)\left[(n+1)^3 - 1 - n(2n+1) - 2n\right]}{4}$$

$$= \frac{(n+1)(n^3 + 3n^2 + 3n + 1 - 1 - 2n^2 - n - 2n)}{4}$$

$$= \frac{(n+1)(n^3 + n^2)}{4}$$

$$= \frac{n^2(n+1)^2}{4} \quad \checkmark$$

> Another challenging 3-mark proof with links to the previous parts of the question. The reason the examiner provides the answers in 'show that' questions is to give you a goal to aim for. Even if you don't achieve the result, you can still use the answer to solve the next part of the question (so you can't be penalised twice for the same mistake).

WORKED SOLUTIONS

Question 16 (15 marks)

a   i   RTP: $\underset{\sim}{v} = ue^{-t}\underset{\sim}{i} + \left(ge^{-t} - g\right)\underset{\sim}{j}$

satisfies $\dot{\underset{\sim}{v}} = -\dot{x}\underset{\sim}{i} - \left(\dot{y} + g\right)\underset{\sim}{j}$.

Proof:

Separate the components and differentiate with respect to t:

$\dot{x} = ue^{-t},$ $\dot{y} = ge^{-t} - g$

$\ddot{x} = -ue^{-t}$ $\ddot{y} = -ge^{-t}$ ✓

In vector form, acceleration $\dot{\underset{\sim}{v}} = \ddot{x}\underset{\sim}{i} + \ddot{y}\underset{\sim}{j}$

Substituting $\dot{\underset{\sim}{v}} = (-ue^{-t})\underset{\sim}{i} + (-ge^{-t})\underset{\sim}{j}$,

now $ue^{-t} = \dot{x}$ and $ge^{-t} = \dot{y} + g$. ✓

So in terms of the components of velocity:

$\dot{\underset{\sim}{v}} = (-\dot{x})\underset{\sim}{i} + \left[-(\dot{y} + g)\right]\underset{\sim}{j}$

$= -\dot{x}\underset{\sim}{i} - (\dot{y} + g)\underset{\sim}{j}$, as required. ✓

> When deriving equations of motion, it is very important to identify the variables (x, v, t) and constants (m, g, e), and be aware of which variable to differentiate with respect to. Keeping the horizontal and vertical components separate is essential. The positive x-direction should always be the same as the direction of motion.

ii   Initial conditions, $t = 0$, $x = 0$, $y = 0$:

$\underset{\sim}{v} = ue^{-t}\underset{\sim}{i} + (ge^{-t} - g)\underset{\sim}{j}$ so separating the components and integrating:

$\dot{x} = ue^{-t}$ $\dot{y} = ge^{-t} - g$

$x = -ue^{-t} + c_1$ ✓ $y = -ge^{-t} - gt + c_2$

$0 = -ue^{-0} + c_1$ $0 = -ge^{-0} - g(0) + c_2$

$c_1 = u$ $c_2 = g$

$\therefore x = ue^{-t} + u$ $y = -ge^{-t} - gt + g$

$\therefore x = u(1 - e^{-t})$ $y = g(1 - t - e^{-t})$

✓   ✓

> By separating components, we see that it is fairly similar to a Maths Extension 1 projectile motion question.

iii   For the bullet to reach the target, $x = R$ and $y = -H$:

$R = u(1 - e^{-t})$

$-H = g(1 - t - e^{-t})$ ✓

$\dfrac{R}{u} = 1 - e^{-t} \Rightarrow -H = g\left(\dfrac{R}{u} - t\right)$

$\therefore -\dfrac{H}{g} = \dfrac{R}{u} - t$

$\dfrac{R}{u} = t - \dfrac{H}{g}$

$= \dfrac{tg - H}{g}$

$\dfrac{u}{R} = \dfrac{g}{tg - H}$

$u = \dfrac{Rg}{gt - H}$ ✓

> Be careful with the signs since the positive direction in the vertical is up.

b   i $\alpha = \cos\theta + i\sin\theta$

$\alpha^k + \alpha^{-k}$

$= (\cos\theta + i\sin\theta)^k + (\cos\theta + i\sin\theta)^{-k}$

$= \cos k\theta + i\sin k\theta + \cos(-k\theta) + i\sin(-k\theta)$

$= \cos k\theta + i\sin k\theta + \cos k\theta - i\sin k\theta$

$= 2\cos k\theta$ ✓

> A common HSC exam question. This question was the last question in the 2017 exam, which was generally well done.

ii $C = \alpha^{-n} + \cdots + \alpha^{-1} + 1 + \alpha + \cdots + \alpha^{n}$

$\qquad = (\alpha^{-1} + \alpha^{-2} + \cdots + \alpha^{-n}) + (1 + \alpha + \cdots + \alpha^{n})$

This is the sum of 2 geometric series, where
$a = \alpha^{-1}, r = \alpha^{-1}, n$ terms (first series)
and $a = 1, r = \alpha, (n + 1)$ terms (second series). ✓

$$C = \frac{\alpha^{-1}(1 - [\alpha^{-1}]^{n})}{1 - \alpha^{-1}} + \frac{1(1 - \alpha^{n+1})}{1 - \alpha} \quad ✓$$

$$= \frac{\alpha^{-1} - \alpha^{-n-1}}{1 - \alpha^{-1}} + \frac{1 - \alpha^{n+1}}{1 - \alpha}$$

$$= \frac{(1 - \alpha)(\alpha^{-1} - \alpha^{-n-1}) + (1 - \alpha^{-1})(1 - \alpha^{n+1})}{(1 - \alpha^{-1})(1 - \alpha)}$$

$$= \frac{\alpha^{-1} - \alpha^{-n-1} - 1 + \alpha^{-n} + 1 - \alpha^{n+1} - \alpha^{-1} + \alpha^{n}}{(1 - \bar{\alpha})(1 - \alpha)} \quad \text{since } \bar{\alpha} = \alpha^{-1}$$

$$= \frac{\alpha^{n} + \alpha^{-n} - \alpha^{n+1} - \alpha^{-n-1}}{(1 - \bar{\alpha})(1 - \alpha)}$$

$$\therefore C = \frac{\alpha^{n} + \alpha^{-n} - (\alpha^{n+1} + \alpha^{-(n+1)})}{(1 - \alpha)(1 - \bar{\alpha})} \quad \text{QED.} \ ✓$$

With careful observation, we see that the series is a sum of 2 different series. Look at the clue in the denominator of the result. Also, it is worth noting that if $\alpha = \cos\theta + i\sin\theta$ then $\bar{\alpha} = \dfrac{1}{\alpha}$.
A common error was not knowing whether the number of terms in each series was n or $n + 1$.

iii $\alpha^{-n} + \cdots + \alpha^{-1} + 1 + \alpha + \cdots + \alpha^{n} = \dfrac{\alpha^{n} + \alpha^{-n} - (\alpha^{n+1} + \alpha^{-(n+1)})}{(1 - \alpha)(1 - \bar{\alpha})}$

Substituting $\alpha = \cos\theta + i\sin\theta$:

$(\cos\theta + i\sin\theta)^{-n} + \cdots + (\cos\theta + i\sin\theta)^{-1} + 1 + (\cos\theta + i\sin\theta) + (\cos\theta + i\sin\theta)^{2} + \cdots + (\cos\theta + i\sin\theta)^{n}$

$$= \frac{(\cos\theta + i\sin\theta)^{n} + (\cos\theta + i\sin\theta)^{-n} - \left[(\cos\theta + i\sin\theta)^{n+1} + (\cos\theta + i\sin\theta)^{-(n+1)}\right]}{\left[1 - (\cos\theta + i\sin\theta)\right]\left[1 - (\cos\theta - i\sin\theta)\right]} \quad ✓$$

$\therefore \cos n\theta \cancel{- i\sin n\theta} + \cdots + \cos\theta \cancel{- i\sin\theta} + 1 + \cos\theta + \cancel{i\sin\theta} + \cdots + \cos n\theta + \cancel{i\sin n\theta}$

$$= \frac{\cos n\theta + i\sin n\theta + \cos(-n\theta) + i\sin(-n\theta) - \cos(n+1)\theta - i\sin(n+1)\theta - \cos[-(n+1)]\theta - i\sin[-(n+1)]\theta}{1 - (\cos\theta - i\sin\theta) - (\cos\theta + i\sin\theta) + (\cos^{2}\theta - i^{2}\sin^{2}\theta)}$$

$\therefore 1 + 2(\cos\theta + \cos 2\theta + \cdots + \cos n\theta)$

$$= \frac{\cos n\theta + i\sin n\theta + \cos n\theta - i\sin n\theta - \cos(n+1)\theta - i\sin(n+1)\theta - \cos(n+1)\theta + i\sin(n+1)\theta}{1 - 2\cos\theta + 1}$$

$$= \frac{2\cos n\theta - 2\cos(n+1)\theta}{2 - 2\cos\theta}$$

$$= \frac{\cos n\theta - \cos(n+1)\theta}{1 - \cos\theta}$$

$$\therefore 1 + 2(\cos\theta + \cos 2\theta + \cos 3\theta + \ldots + \cos n\theta) = \frac{\cos n\theta - \cos(n+1)\theta}{1 - \cos\theta} \quad \text{QED.} \ ✓$$

Note that the complexity in many Maths Extension 2 questions is in the algebra rather than the concepts. Your algebra skills must be perfect and precise in order to arrive at the result.

iv Using the identity above, let $\theta = \dfrac{\pi}{n}$.

$$1 + 2\left[\cos\frac{\pi}{n} + \cos 2\left(\frac{\pi}{n}\right) + \cos 3\left(\frac{\pi}{n}\right) + \ldots + \cos n\left(\frac{\pi}{n}\right)\right] = \frac{\cos n\left(\frac{\pi}{n}\right) - \cos(n+1)\left(\frac{\pi}{n}\right)}{1 - \cos\frac{\pi}{n}}$$

$$2\left[\cos\frac{\pi}{n} + \cos 2\left(\frac{\pi}{n}\right) + \cos 3\left(\frac{\pi}{n}\right) + \ldots + \cos n\left(\frac{\pi}{n}\right)\right] = \frac{\cos n\left(\frac{\pi}{n}\right) - \cos(n+1)\left(\frac{\pi}{n}\right)}{1 - \cos\frac{\pi}{n}} - 1$$

$$\cos\frac{\pi}{n} + \cos\frac{2\pi}{n} + \cos\frac{3\pi}{n} + \ldots + \cos\pi = \frac{\cos\pi - \cos(n+1)\left(\frac{\pi}{n}\right) - \left(1 - \cos\frac{\pi}{n}\right)}{2\left(1 - \cos\frac{\pi}{n}\right)}$$

$$= \frac{-1 - \cos\left(\pi + \frac{\pi}{n}\right) - 1 + \cos\frac{\pi}{n}}{2\left(1 - \cos\frac{\pi}{n}\right)}$$

$$= \frac{-1 - -\cos\frac{\pi}{n} - 1 + \cos\frac{\pi}{n}}{2\left(1 - \cos\frac{\pi}{n}\right)}$$

$$= \frac{-2 + 2\cos\frac{\pi}{n}}{2\left(1 - \cos\frac{\pi}{n}\right)}$$

$$= \frac{-2\left(1 - \cos\frac{\pi}{n}\right)}{2\left(1 - \cos\frac{\pi}{n}\right)}$$

$$= -1, \text{ independent of } n \ \checkmark$$

A challenging question overall where all the parts were related. Part **iv** requires strong skills in algebra and trigonometric identities to simplify the expression. Realising that substituting the appropriate value for θ will lead to the result is the mathematical insight required in this final part.

The 2020 Mathematics Extension 2 HSC Exam Worked Solutions

The 2020 HSC exam and other past HSC papers can be downloaded from the NESA website (www.educationstandards.nsw.edu.au) by selecting 'Year 11 – Year 12', 'HSC exam papers'. NESA marking feedback and guidelines can also be found there.

Breakdown of topics by marks in the 2020 HSC exam:

1. Proof   18%
2. 3D vectors   15%
3. Complex numbers   21%
4. Further integration   16%
5. Mechanics   30%

Section I (1 mark each)

Question 1

B $\text{Length} = \sqrt{(-1)^2 + 18^2 + (-6)^2}$

$= \sqrt{361}$

$= 19$

Make sure you take the square root to find the magnitude of the vector.

Question 2

B $(3 + i)^2 + p(3 + i) + q = 0 + 0i$

$9 + 6i + i^2 + 3p + pi + q = 0 + 0i$

$9 + (6 + p)i - 1 + 3p + q = 0 + 0i$

$(8 + 3p + q) + (6 + p)i = 0 + 0i$

Equating real and imaginary:

$8 + 3p + q = 0 \quad 6 + p = 0$

$p = -6$

$8 + 3(-6) + q = 0$

$8 - 18 + q = 0$

$-10 + q = 0$

$q = 10$

A typical exam question. Once terms are expanded and collected, you need to equate real and imaginary components.

Question 3

C $x = 1 - 2\lambda$ and $y = 3 + 4\lambda$

Eliminating λ:

$2\lambda = 1 - x$

$y = 3 + 2(1 - x)$

$= 3 + 2 - 2x$

$= -2x + 5$

$y + 2x = 5$

This is a 2D Cartesian equation.

Remember, $\underset{\sim}{r} = \begin{pmatrix} x \\ y \end{pmatrix}$.

Question 4

A Let $z = re^{i\theta}$ [or $r(\cos\theta + i\sin\theta)$]

$z^2 = (re^{i\theta})^2$

$= r^2 e^{2i\theta}$

$|z| = r$

$\dfrac{z^2}{|z|} = \dfrac{r^2 e^{2i\theta}}{r} = re^{2i\theta}$

So same modulus, double the argument of z: diagram A.

In exponential or polar form, it is easy to see that $\dfrac{z^2}{|z|}$ has the same modulus and double the argument of z.

Question 5

D Max $v = na = 4$ [1] and Max $\ddot{x} = n^2 a = 6$ [2]

[2] ÷ [1]: $\quad n = \dfrac{6}{4} = \dfrac{3}{2}$

$\text{Period} = \dfrac{2\pi}{n}$

$= \dfrac{2\pi}{\frac{3}{2}}$

$= \dfrac{4\pi}{3}$

Maximum acceleration is at the extremes and maximum velocity is as it passes through the centre of motion.

Question 6

A $\displaystyle\int \dfrac{1}{x^2 + 4x + 10}\,dx = \int \dfrac{1}{(x^2 + 4x + 4) + 6}\,dx$

$= \displaystyle\int \dfrac{1}{(x + 2)^2 + (\sqrt{6})^2}\,dx$

$= \dfrac{1}{\sqrt{6}} \tan^{-1}\left(\dfrac{x + 2}{\sqrt{6}}\right) + c$

Complete the square in the denominator and use

$\displaystyle\int \dfrac{1}{a^2 + x^2}\,dx = \dfrac{1}{a}\tan^{-1}\left(\dfrac{x}{a}\right) + c.$

A typical integration question.

Question 7

D 'If $2^n - 1$ is not prime, then n is not prime.'

A statement that would disprove this proposition would be a counterexample where $2^n - 1$ is not prime, but n is prime.

Statements B and D refer to $2^n - 1$ not being prime (being divisible by numbers other than 1 and itself), but only D has $n = 11$ being prime.

Any statement can be disproved by one counterexample.

9780170459273

Question 8

B Negation of statement:

'If n is even, then if n is a multiple of 3, then n is a multiple of 6'

Negation: 'n is even, and is a multiple of 3, but not a multiple of 6'.

> The negation of a statement of the form 'If P, then Q' can be created by changing it to 'If P, then not Q'.
> The most popular answer (D) was incorrect, where NOT appeared twice.
> Questions 7 and 8 test the language and structure of proof.

Question 9

C $z = e^{i\theta}$ is a complex number with a modulus of 1, so it lies on the unit circle in the complex plane.

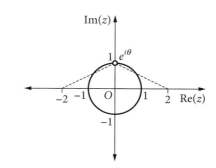

$\left|e^{i\theta} - 2\right|$ is the length of the vector from 2 to z, and $\left|e^{i\theta} + 2\right|$ is the length of the vector from -2 to z.

The maximum value of $\left|e^{i\theta} - 2\right| + \left|e^{i\theta} + 2\right|$ occurs when z is at the top (or bottom) of the circle at $z = 1$ (or -1).

$\left|e^{i\theta} - 2\right| = \sqrt{1^2 + 2^2} = \sqrt{5}$ and $\left|e^{i\theta} + 2\right| = \sqrt{1^2 + 2^2} = \sqrt{5}$

The maximum value is $\sqrt{5} + \sqrt{5} = 2\sqrt{5}$.

> A diagram with this type of problem is always useful to understand the question better.
> This problem can also be solved algebraically in polar form, but it requires more working.

Question 10

B $\int_0^a f(x) + f(2a - x)\,dx$

$= \int_0^a f(x)\,dx + \int_0^a f(2a - x)\,dx$

$= \int_0^a f(x)\,dx + \int_{2a}^a f(u)(-du)$ (substituting $u = 2a - x$, $du = -dx$ and changing limits)

$= \int_0^a f(x)\,dx - \int_{2a}^a f(u)\,du$

$= \int_0^a f(x)\,dx + \int_a^{2a} f(u)\,du$ (swapping the limits and 'negating the integral')

$= \int_0^a f(x)\,dx + \int_a^{2a} f(x)\,dx$ (changing the dummy variable back to x and combining integrals over the 2 adjacent domains)

$= \int_0^{2a} f(x)\,dx$

> This question tests deep conceptual understanding of integration.
> Do not forget to change the limits and remember, $\int_a^{-a} f(x)\,dx = -\int_{-a}^a f(x)\,dx$.

Section II ($\checkmark$ = 1 mark)

Question 11

a **i** $|w| = \sqrt{(-1)^2 + 4^2}$

       $= \sqrt{17}$ $\checkmark$

   ii $w\bar{z} = (-1 + 4i)(2 + i)$ $\checkmark$

         $= -2 - i + 8i - 4$

         $= -6 + 7i$ $\checkmark$

> Fairly straightforward and common complex number operations.

b Using integration by parts:

Let $u = \ln x$, $v' = x$.

Hence, $u' = \dfrac{1}{x}$ and $v = \dfrac{1}{2}x^2$. $\checkmark$

$\displaystyle\int_1^e x\ln(x)\,dx = \left[\frac{1}{2}x^2\ln x\right]_1^e - \frac{1}{2}\int_1^e x\,dx$

$\qquad\qquad = \dfrac{1}{2}e^2 - \dfrac{1}{2}\left[\dfrac{x^2}{2}\right]_1^e$

$\qquad\qquad = \dfrac{1}{2}e^2 - \dfrac{1}{2}\left(\dfrac{e^2}{2} - \dfrac{1}{2}\right)$ $\checkmark$

$\qquad\qquad = \dfrac{1}{4}e^2 + \dfrac{1}{4}$ $\checkmark$

$\qquad\qquad = \dfrac{1}{4}(e^2 + 1)$

> Standard application of integration by parts. It is imperative to choose $u = \ln x$ to be able to make a start.

c $a = v^2 + v$

$v\dfrac{dv}{dx} = v(v + 1)$ $\checkmark$

$\dfrac{dv}{v + 1} = dx$

Integrating:

$\displaystyle\int \frac{dv}{v + 1} = \int dx$

$x = \ln|v + 1| + c$ $\checkmark$

When $x = 0$, $v = 1$:

$0 = \ln(1 + 1) + c$

$c = -\ln 2$

$x = \ln|v + 1| - \ln 2$ $\checkmark$

$x = \ln\dfrac{|v + 1|}{2}$

> Remember to choose $a = v\dfrac{dv}{dx}$ in this application as we need x as a function of v.

d $\underset{\sim}{u} = -2\underset{\sim}{i} - \underset{\sim}{j} + 3\underset{\sim}{k}$ $\underset{\sim}{v} = p\underset{\sim}{i} + \underset{\sim}{j} + 2\underset{\sim}{k}$

$\underset{\sim}{u} - \underset{\sim}{v} = -2\underset{\sim}{i} - \underset{\sim}{j} + 3\underset{\sim}{k} - p\underset{\sim}{i} - \underset{\sim}{j} - 2\underset{\sim}{k}$

$\qquad = (-2 - p)\underset{\sim}{i} - 2\underset{\sim}{j} + \underset{\sim}{k}$ $\checkmark$

$\underset{\sim}{u} + \underset{\sim}{v} = -2\underset{\sim}{i} - \underset{\sim}{j} + 3\underset{\sim}{k} + p\underset{\sim}{i} + \underset{\sim}{j} + 2\underset{\sim}{k}$

$\qquad = (p - 2)\underset{\sim}{i} + 0\underset{\sim}{j} + 5\underset{\sim}{k}$

If perpendicular, $(\underset{\sim}{u} - \underset{\sim}{v}) \cdot (\underset{\sim}{u} + \underset{\sim}{v}) = 0$

$-(2 + p)(p - 2) - 2(0) + 5(1) = 0$ $\checkmark$

$\qquad\qquad -(p^2 - 4) + 5 = 0$

$\qquad\qquad\qquad\qquad p^2 = 9$

$\qquad\qquad\qquad\qquad p = \pm 3$ $\checkmark$

> Expect the relationship between perpendicular vectors and the scalar product to be tested often.

e $z^2 + 3z + (3 - i) = 0$

$z = \dfrac{-b \pm \sqrt{b^2 - 4ac}}{2a}$

$\quad = \dfrac{-3 \pm \sqrt{3^2 - 4(1)(3 - i)}}{2(1)}$

$\quad = \dfrac{-3 \pm \sqrt{9 - 12 + 4i}}{2}$

$\quad = \dfrac{-3 \pm \sqrt{-3 + 4i}}{2}$ $\checkmark$

Now solve:

$\sqrt{-3 + 4i} = a + ib$, where $a, b \in \mathbb{R}$

$-3 + 4i = a^2 - b^2 + 2abi$

Equating real and imaginary parts:

$-3 = a^2 - b^2$, $2 = ab$ $\checkmark$

By inspection, $a = 1, b = 2$ or $a = -1, b = -2$.

$\therefore \sqrt{-3 + 4i} = \pm(1 + 2i)$ $\checkmark$

$\therefore$ For $z^2 + 3z + (3 - i) = 0$

$z = \dfrac{-3 \pm (1 + 2i)}{2}$

$\quad = \dfrac{-2 + 2i}{2}$ or $\dfrac{-4 - 2i}{2}$

$\quad = -1 + i$ or $-2 - i$ $\checkmark$

> This 4-mark question has no scaffolding (clues), so you are expected to find the square root of a complex number without guidance. 4 marks means that this lengthy process has 4 main steps that you need to show. There are many methods for solving this quadratic equation, but the above method is the most efficient.

Question 12

a i Resolving forces vertically:

$$R + 200 \sin 30° = 50g \checkmark$$

$$R + 200 \times \frac{1}{2} = 50 \times 10$$

$$R + 100 = 500$$

$$R = 400 \text{ newtons} \checkmark$$

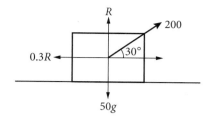

A force diagram will assist in choosing the correct trigonometric ratio, choosing 'up' as the positive direction. The best students know how to resolve vertical and horizontal forces easily. Some students mistakenly read $50g$ as '50 grams' rather than 50 × gravitational acceleration.

ii $m\ddot{x} = 200 \cos 30° - 0.3R \checkmark$

$$= \frac{200\sqrt{3}}{2} - 0.3 \times 400$$

$$= 100\sqrt{3} - 120$$

$$= 53.2050\ldots$$

$$\approx 53.2 \text{ newtons} \checkmark$$

Be skilled at writing a force equation, in this case, the horizontal net force, $m\ddot{x} = \sum F_x$, choosing the positive direction to the right. Show enough working for a 'show that' question.

iii $50\ddot{x} = 53.2$

$$\ddot{x} = \frac{53.2}{50} = 1.064$$

$$\frac{dv}{dt} = 1.064 \checkmark$$

$$v = 1.064t + c$$

When $t = 0$, $v = 0$, $\therefore c = 0$.

$v = 1.064t$

Velocity after 3 seconds:

$$v = 1.064 \times 3$$

$$= 3.192 \text{ m s}^{-1} \checkmark$$

Generally well done. Part **iii** uses the result from part **ii**, as well as $m = 50$. You should notice that the motion is all horizontal, no vertical. If v as a function of t is needed, choose $\dfrac{dv}{dt}$ for acceleration. Noting that velocity is the integral of acceleration we could also use $\int_0^v dv = \int_0^3 1.064 \, dt$, which gives $v = [1.064t]_0^3 = 3.192 \text{ ms}^{-1}$.

b i $\underset{\sim}{a}(t) = \begin{pmatrix} 0 \\ -g \end{pmatrix}$

$$\underset{\sim}{v}(t) = \begin{pmatrix} 0 + c_1 \\ -gt + c_2 \end{pmatrix}$$

$$\underset{\sim}{v}(0) = \begin{pmatrix} c_1 = u\cos\theta \\ c_2 = u\sin\theta \end{pmatrix}$$

Hence, $\underset{\sim}{v}(t) = \begin{pmatrix} 0 + u\cos\theta \\ -gt + u\sin\theta \end{pmatrix}$. ✓

$$\underset{\sim}{r}(t) = \begin{pmatrix} ut\cos\theta + c_3 \\ -\dfrac{1}{2}gt^2 + ut\sin\theta + c_4 \end{pmatrix}$$

$$\underset{\sim}{r}(0) = \begin{pmatrix} c_3 = 0 \\ c_4 = 0 \end{pmatrix}$$ ✓

Hence, $\underset{\sim}{r}(t) = \begin{pmatrix} ut\cos\theta \\ ut\sin\theta - \dfrac{1}{2}gt^2 \end{pmatrix}$, as required. ✓

A common HSC exam question (also in Maths Extension 1): deriving the equations of motion for projectile motion. Learn how to set out the proof correctly, especially for a 'show that' question. Many students just wrote memorised formulas here. Start with the information provided, integrate components using Cartesian or vector notation and find constants for velocity components and then displacement components, showing working. Remember to use the mark value (3 marks) to guide your working.

ii $x = ut\cos\theta \rightarrow t = \dfrac{x}{u\cos\theta}$

$y = ut\sin\theta - \dfrac{1}{2}gt^2$

Sub t into y-equation:

$y = u\left(\dfrac{x}{u\cos\theta}\right)\sin\theta - \dfrac{1}{2}g\left(\dfrac{x}{u\cos\theta}\right)^2$ ✓

$\qquad = x\tan\theta - \dfrac{gx^2}{2u^2}(\sec\theta)^2$ ✓

$\qquad = x\tan\theta - \dfrac{gx^2}{2u^2}\left(1 + \tan^2\theta\right)$

$\qquad = -\dfrac{gx^2}{2u^2}\left(-\dfrac{2u^2}{gx}\tan\theta + 1 + \tan^2\theta\right)$

$\qquad = -\dfrac{gx^2}{2u^2}\left(\tan^2\theta - \dfrac{2u^2}{gx}\tan\theta + 1\right)$ ✓

Another common HSC exam question that has been moved to the Maths Extension 2 course from Extension 1: the path of a projectile.

Remember to use t as a parameter and eliminate t from x and y expressions.

Use $\sec^2\theta = 1 + \tan^2\theta$ to eliminate the $\sec^2\theta$, and be careful with the signs when simplifying to reach the required form. Again, a lot of intricate algebra where it's very easy for students to make careless errors or lose track. Look at the equation given in the question that you are aiming for. The HSC examiners remind students to write clearly and carefully, keeping track of squares and negative signs.

iii When $x = R$, this quadratic equation in $\tan\theta$ has 2 distinct values.

$$0 = -\frac{gR^2}{2u^2}\left(\tan^2\theta - \frac{2u^2}{gR}\tan\theta + 1\right)$$

$\therefore \tan^2\theta - \frac{2u^2}{gR}\tan\theta + 1 = 0$ and $\Delta > 0$

for 2 distinct values.

$$\Delta = b^2 - 4ac$$

$$= \left(-\frac{2u^2}{gR}\right)^2 - 4$$

$$= \frac{4u^4}{g^2R^2} - 4 \quad \checkmark$$

$$= \frac{4u^4 - 4g^2R^2}{g^2R^2}$$

$$= \frac{4(u^4 - g^2R^2)}{g^2R^2}$$

As $u^2 > gR > 0$, $u^4 > g^2R^2 > 0$

$\therefore u^4 - g^2R^2 > 0$

$\therefore \Delta = \frac{4(u^4 - g^2R^2)}{g^2R^2} > 0.$

So the quadratic equation in $\tan\theta$ has 2 distinct values for which the particle will land at $x = R$. $\checkmark$

Note that for 2 values of θ, the quadratic in $\tan\theta$ must have 2 solutions and a discriminant greater than zero. The discriminant was studied at the beginning of Year 11. Note also that the question does not ask you to solve the equation, which some HSC students tried to do. Be very careful when using the given condition $u^2 > gR$ in proving $\Delta > 0$.

Question 13

a From the HSC exam reference sheet, the displacement equation for simple harmonic motion (SHM) is:

$$x = a\cos(nt + \alpha) + c$$

$$\text{Period} = \frac{2\pi}{n} = \frac{\pi}{3}$$

$$n = \frac{6\pi}{\pi} = 6 \quad \checkmark$$

$a = $ amplitude $= 2\sqrt{3}$ $\checkmark$

$c = $ centre $= \sqrt{3}$

$x = 2\sqrt{3}\cos(6t + \alpha) + \sqrt{3}$

When $t = 0$, $x = 3\sqrt{3}$ at endpoint:

$$3\sqrt{3} = 2\sqrt{3}\cos(\alpha) + \sqrt{3}$$

$$2\sqrt{3} = 2\sqrt{3}\cos(\alpha)$$

$$1 = \cos\alpha$$

$$\alpha = 0$$

So $x = 2\sqrt{3}\cos(6t) + \sqrt{3}$. $\checkmark$

Simple harmonic motion is another topic that has moved from Maths Extension 1 to Maths Extension 2. Students who use the HSC exam reference sheet formula $x = a\cos(nt + \alpha) + c$ and draw a diagram are much more likely to make progress with this question. It's better to choose the cosine function rather than sine when the motion starts at an extreme point, not the centre. In SHM questions, an understanding of the physical relationship with the equations is a much better approach, for example, knowing that speed is a maximum when the particle is at the centre.

b $\underset{\sim}{r} = \begin{pmatrix} 3 \\ -1 \\ 7 \end{pmatrix} + \lambda_1 \begin{pmatrix} 1 \\ 2 \\ 1 \end{pmatrix}$ and $\underset{\sim}{r} = \begin{pmatrix} 3 \\ -6 \\ 2 \end{pmatrix} + \lambda_2 \begin{pmatrix} -2 \\ 1 \\ 3 \end{pmatrix}$

Equating coordinates:

$$\begin{pmatrix} 3 \\ -1 \\ 7 \end{pmatrix} + \lambda_1 \begin{pmatrix} 1 \\ 2 \\ 1 \end{pmatrix} = \begin{pmatrix} 3 \\ -6 \\ 2 \end{pmatrix} + \lambda_2 \begin{pmatrix} -2 \\ 1 \\ 3 \end{pmatrix} \checkmark$$

$3 + \lambda_1 = 3 - 2\lambda_2$ [1]

$-1 + 2\lambda_1 = -6 + \lambda_2$ [2]

$7 + \lambda_1 = 2 + 3\lambda_2$ [3]

Simplifying:

From [1]: $\lambda_1 = -2\lambda_2$

Substitute into [3]:

$$7 + (-2\lambda_2) = 2 + 3\lambda_2$$
$$5 = 5\lambda_2$$
$$\lambda_2 = 1 \ \checkmark$$

$$\therefore \lambda_1 = -2 \times 1 = -2$$

Check in [2]: $-1 + 2(-2) = -6 + 1 = -5$, true.

Substituting:

Point is

$x = 3 + (-2) = 1$

$y = -1 + 2(-2) = -5$

$z = 7 + (-2) = 5.$

$(1, -5, 5) \ \checkmark$

Most students could complete this question, though some thought they finished after finding λ_1 and λ_2. Equating the components and solving for the λs to satisfy all 3 equations is necessary.

Answer the whole question by stating the point of intersection. The HSC markers remind you to check your solution.

c i RTP: $\dfrac{a+b}{2} \geq \sqrt{ab}$

Using the triangle:

$(a+b)^2 = x^2 + (a-b)^2$ (Pythagoras' theorem)

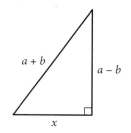

$a^2 + 2ab + b^2 = x^2 + a^2 - 2ab + b^2$

$\qquad 2ab = x^2 - 2ab$

$\qquad\quad x^2 = 4ab$

$\qquad\quad\ x = 2\sqrt{ab}$ ✓

Also, $x \leq a+b$ ($a+b$ is the hypotenuse).

$2\sqrt{ab} \leq a+b$

$\ \sqrt{ab} \leq \dfrac{a+b}{2}$ ✓

OR

Consider the difference:

$\dfrac{a+b}{2} - \sqrt{ab} = \dfrac{a+b-2\sqrt{ab}}{2}$

$\qquad\qquad\qquad = \dfrac{\left(\sqrt{a}\right)^2 - 2\sqrt{a}\sqrt{b} + \left(\sqrt{b}\right)^2}{2}$

$\qquad\qquad\qquad = \dfrac{\left(\sqrt{a} - \sqrt{b}\right)^2}{2}$ ✓

$\qquad\qquad\qquad \geq 0 \ \text{ since } \left(\sqrt{a} - \sqrt{b}\right)^2 \geq 0$

Therefore, $\dfrac{a+b}{2} \geq \sqrt{ab}$. ✓

This common proof demonstrates the importance of knowing basic Years 9 and 10 algebraic skills such as expanding perfect squares, $(a+b)^2$. What was unusual was the use of a right-angled triangle, as this proof is usually done completely algebraically. This is the famous arithmetic mean $\left(\dfrac{a+b}{2}\right)$ / geometric mean $\left(\sqrt{ab}\right)$ (AM-GM) inequality that forms the basis of almost all inequality proof questions in the Maths Extension 2 course. The HSC markers recommend that you learn rather than memorise this proof. Correct setting out is essential to the argument.

ii RTP: $p^2 + 4q^2 \geq 4pq$

Proof: We know from part **i** that $\dfrac{a+b}{2} \geq \sqrt{ab}$ or $a+b \geq 2\sqrt{ab}$ when $a > b \geq 0$. By exchanging a and b, we can see that it is also true when $b > a \geq 0$, and when $a = b$ the 2 sides are equal. So the result is true for any non-negative a and b.

Let $a = p^2$ and $b = 4q^2$:

$p^2 + 4q^2 \geq 2\sqrt{p^2 4q^2}$

$p^2 + 4q^2 \geq 2p2q$

$p^2 + 4q^2 \geq 4pq$ ✓

Using an appropriate substitution in part **i** easily leads to this result. Always look for a connection between the parts. The result could also be proved by expanding and rearranging the equation $(p-2q)^2 \geq 0$. Note that this question is only worth one mark, which indicates that the proof should not be long and complicated.

d i   RTP: $e^{in\theta} + e^{-in\theta} = 2\cos n\theta$

Using De Moivre's theorem:

$$e^{in\theta} + e^{-in\theta} = \cos n\theta + i\sin n\theta + \cos(-n\theta) + i\sin(-n\theta)$$
$$= \cos n\theta + i\sin n\theta + \cos(n\theta) - i\sin(n\theta) \text{ as } \sin(-n\theta) = -\sin(n\theta)$$
$$= 2\cos(n\theta) \checkmark$$

This is a 'FAQ' in the HSC exam. Again, only worth 1 mark, and most students did well and showed sufficient working.

ii   Using the binomial theorem:

$$\left(e^{i\theta} + e^{-i\theta}\right)^4 = \left(e^{i\theta}\right)^4 + 4\left(e^{i\theta}\right)^3\left(e^{-i\theta}\right)^1 + 6\left(e^{i\theta}\right)^2\left(e^{-i\theta}\right)^2 + 4\left(e^{i\theta}\right)^1\left(e^{-i\theta}\right)^3 + \left(e^{-i\theta}\right)^4 \checkmark$$

$$(2\cos\theta)^4 = e^{i4\theta} + 4e^{i2\theta} + 6e^0 + 4e^{-i2\theta} + e^{-i4\theta}$$
$$= \left(e^{i4\theta} + e^{-i4\theta}\right) + 4\left(e^{i2\theta} + e^{-i2\theta}\right) + 6 \quad \text{(using part i)} \checkmark$$
$$= 2\cos 4\theta + 4(2\cos 2\theta) + 6$$
$$16\cos^4\theta = 2\cos 4\theta + 8\cos 2\theta + 6$$
$$\cos^4\theta = \frac{1}{8}\cos 4\theta + \frac{1}{2}\cos 2\theta + \frac{3}{8}$$
$$\cos^4\theta = \frac{1}{8}(\cos 4\theta + 4\cos 2\theta + 3) \checkmark$$

Always look at your answer to part **i** as a clue, even if the word 'hence' is absent. This question demonstrates the importance of knowing the binomial theorem, a Year 11 Maths Extension 1 concept. This is a common application of the expansion of $(e^{i\theta} + e^{-i\theta})^n$. Note the grouping of the symmetrical pairs.

iii $\int_0^{\frac{\pi}{2}} \cos^4\theta\, d\theta = \int_0^{\frac{\pi}{2}} \frac{1}{8}(\cos 4\theta + 4\cos 2\theta + 3)\, d\theta$

$$= \frac{1}{8}\left[\frac{1}{4}\sin 4\theta + 2\sin 2\theta + 3\theta\right]_0^{\frac{\pi}{2}} \checkmark$$

$$= \frac{1}{8}\left[\frac{1}{4}\sin 4\left(\frac{\pi}{2}\right) + 2\sin 2\left(\frac{\pi}{2}\right) + 3\left(\frac{\pi}{2}\right) - \left(\frac{1}{4}\sin 0 + 2\sin 0 + 0\right)\right]$$

$$= \frac{1}{8}\left(\frac{1}{4}\sin 2\pi + 2\sin \pi + \frac{3\pi}{2} - 0\right)$$

$$= \frac{1}{8}\left(0 + 0 + \frac{3\pi}{2}\right)$$

$$= \frac{3\pi}{16} \checkmark$$

The 'Hence' in this question means to use the answer to part **ii** to answer part **iii**. The answer was given in part **ii**, a 'show that' question, so that even if you cannot complete part **ii**, you can still use its answer for part **iii**. Always use the identity given in the examination rather than an incorrect one.

Question 14

a **i** Given $z_2 = z_1 e^{\frac{i\pi}{3}}$,

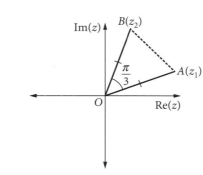

$$|z_2| = \left| z_1 e^{\frac{i\pi}{3}} \right|$$

$$= |z_1| \left| e^{\frac{i\pi}{3}} \right|$$

$$= |z_1| \times 1$$

$$= |z_1|.$$

So $OA = OB$, ΔOAB is isosceles. ✓

$$\arg z_2 = \arg \left(z_1 e^{\frac{i\pi}{3}} \right)$$

$$= \arg(z_1) + \arg \left(e^{\frac{i\pi}{3}} \right)$$

$$= \arg(z_1) + \frac{\pi}{3}$$

$\angle BOA = \dfrac{\pi}{3}$ (or 60°) and $\angle OAB = \angle OBA = \dfrac{1}{2}\left(\pi - \dfrac{\pi}{3} \right) = \dfrac{\pi}{3}$ (or 60°) (angle sum of isosceles triangle)

Therefore, ΔOAB is equilateral (3 equal angles). ✓

> This question links complex numbers with basic triangle geometry. You should know that multiplying by a complex number is a rotation and an isosceles triangle is equilateral if the angle between the equal sides is 60°. This question tests your geometrical understanding of complex numbers in polar/exponential form and your ability to reason and explain in words.

ii RTP: $z_1^2 + z_2^2 = z_1 z_2$.

Proof:

$$\text{LHS} = z_1^2 + \left(z_1 e^{\frac{i\pi}{3}} \right)^2 \qquad\qquad\qquad \text{RHS} = z_1 z_2$$

$$= z_1^2 + z_1^2 e^{\frac{i2\pi}{3}} \qquad\qquad\qquad\qquad\quad = z_1 z_1 e^{\frac{i\pi}{3}}$$

$$= z_1^2 \left(1 + \cos\frac{2\pi}{3} + i\sin\frac{2\pi}{3} \right) \qquad\quad = z_1^2 \left(\cos\frac{\pi}{3} + i\sin\frac{\pi}{3} \right)$$

$$= z_1^2 \left[1 + \left(-\frac{1}{2} \right) + i\frac{\sqrt{3}}{2} \right] \qquad\qquad = z_1^2 \left(\frac{1}{2} + i\frac{\sqrt{3}}{2} \right) ✓$$

$$= z_1^2 \left(\frac{1}{2} + i\frac{\sqrt{3}}{2} \right) ✓$$

Therefore, LHS = RHS, $z_1^2 + z_2^2 = z_1 z_2$. ✓

> Students struggled with this question. Notice the use of part **i**. More capable students knew when to factorise.

b $a = 10(1 - [kv]^2)$

> **Hint**
> We need to find v when $t = 5$, so we need an equation involving v and t. Do not substitute $k = 0.01$ yet, to save writing 0.0001 over and over.

$$\frac{dv}{dt} = 10(1 - k^2v^2)$$

$$\frac{dt}{dv} = \frac{1}{10(1 - k^2v^2)}$$

Separate $\dfrac{1}{1 - k^2v^2}$ into partial fractions:

$$\frac{1}{1 - k^2v^2} = \frac{1}{(1 - kv)(1 + kv)} = \frac{A}{1 - kv} + \frac{B}{1 + kv} \quad \checkmark$$

$$A(1 + kv) + B(1 - kv) = 1$$
$$A + Akv + B - Bkv = 1$$
$$A + B + kv(A - B) = 1$$

Equating coefficients: $A + B = 1$, $A - B = 0$

Add them:

$$2A = 1$$
$$A = \frac{1}{2}$$

$$B = 1 - \frac{1}{2} = \frac{1}{2}$$

$$\therefore \frac{1}{1 - k^2v^2} = \frac{1}{2}\left(\frac{1}{1 - kv} + \frac{1}{1 + kv}\right)$$

$$\frac{dt}{dv} = \frac{1}{10(1 - k^2v^2)}$$

$$= \frac{1}{10}\frac{1}{2}\left(\frac{1}{1 - kv} + \frac{1}{1 + kv}\right)$$

$$= \frac{1}{20}\left(\frac{1}{1 - kv} + \frac{1}{1 + kv}\right)$$

$$t = \frac{1}{20}\int \frac{1}{1 - kv} + \frac{1}{1 + kv}\, dv$$

$$20t = -\frac{1}{k}\ln|1 - kv| + \frac{1}{k}\ln|1 + kv| + c \quad \checkmark$$

When $t = 0$, $v = 0$:

$$0 = -\frac{1}{k}\ln 1 + \frac{1}{k}\ln 1 + c$$

$$0 = 0 + 0 + c$$

$$20t = -\frac{1}{k}\ln|1 - kv| + \frac{1}{k}\ln|1 + kv|$$

$$20kt = -\ln|1 - kv| + \ln|1 + kv|$$

$$\ln\left|\frac{1 + kv}{1 - kv}\right| = 20kt$$

$$\frac{1 + kv}{1 - kv} = \pm e^{20kt} \quad \checkmark$$

By considering the starting conditions $t = 0$, $v = 0$ we can see that LHS = 1, RHS = ± 1, so the positive solution is correct here. (Alternatively, $a > 0$, so $1 - [kv]^2 > 0$, so $[kv]^2 < 1$, so $0 < kv < 1$ as k, $v > 0$, so $\dfrac{1 + kv}{1 - kv} > 0$.)

Substitute $k = 0.01$, $t = 5$ to find v.

$$\frac{1 + 0.01v}{1 - 0.01v} = e^{20(0.01)5}$$

$$\frac{1 + 0.01v}{1 - 0.01v} = e^1$$

$$1 + 0.01v = e(1 - 0.01v)$$

$$= e - 0.01ev$$

$$0.01v + 0.01ev = e - 1$$

$$0.01v(1 + e) = e - 1$$

$$v = \frac{e - 1}{0.01(1 + e)}$$

$$v \approx 46.2 \,\text{m/s} \quad \checkmark$$

> A 4-mark question with no scaffolding (clues), so a lot of planning and working is required. Recognise when to use partial fractions to integrate. Because ultimately we are finding the value of v when $t = 5$, we don't need to make v the subject of the formula. We can substitute into the formula and solve for v. Questions like this can be very tedious and require strong algebraic skills and accuracy. Note the use of the initial conditions when integrating.

9780170459273

WORKED SOLUTIONS

c RTP: $\dfrac{1}{2^2} + \dfrac{1}{3^2} + \dfrac{1}{4^2} + \cdots + \dfrac{1}{n^2} < \dfrac{n-1}{n}, n \geq 2$

Proof:

Let $P(n)$ be the proposition that $\dfrac{1}{2^2} + \dfrac{1}{3^2} + \dfrac{1}{4^2} + \cdots + \dfrac{1}{n^2} < \dfrac{n-1}{n}, n \geq 2, n \in \mathbb{N}$.

Prove $P(2)$ true:

$$\text{LHS} = \dfrac{1}{2^2} \qquad\qquad \text{RHS} = \dfrac{2-1}{2}$$
$$= \dfrac{1}{4} \qquad\qquad\qquad = \dfrac{1}{2}$$

So LHS < RHS.

$P(2)$ is true. ✓

Assume that $P(k)$ is true for some $k \geq 2$, $k \in \mathbb{N}$.

$$\dfrac{1}{2^2} + \dfrac{1}{3^2} + \dfrac{1}{4^2} + \cdots + \dfrac{1}{k^2} < \dfrac{k-1}{k} \quad [*]$$

RTP $P(k+1)$ is true:

$$\dfrac{1}{2^2} + \dfrac{1}{3^2} + \dfrac{1}{4^2} + \cdots + \dfrac{1}{k^2} + \dfrac{1}{(k+1)^2} < \dfrac{k}{k+1}. \checkmark$$

Proof:

Consider the LHS of $P(k+1)$:

$$\dfrac{1}{2^2} + \dfrac{1}{3^2} + \dfrac{1}{4^2} + \cdots + \dfrac{1}{k^2} + \dfrac{1}{(k+1)^2} < \dfrac{k-1}{k} + \dfrac{1}{(k+1)^2} \quad \text{using } [*]$$

$$= \dfrac{(k-1)(k+1)^2 + k}{k(k+1)^2}$$

$$= \dfrac{(k^2-1)(k+1) + k}{k(k+1)^2} \quad \checkmark$$

$$= \dfrac{k^3 + k^2 - k - 1 + k}{k(k+1)^2}$$

$$= \dfrac{k^3 + k^2 - 1}{k(k+1)^2}$$

$$= \dfrac{k^2(k+1)}{k(k+1)^2} - \dfrac{1}{k(k+1)^2}$$

$$< \dfrac{(k+1)k^2}{k(k+1)^2} \quad \text{since} \quad \dfrac{1}{k(k+1)^2} > 0.$$

$$= \dfrac{k}{k+1}, \quad \text{as required.}$$

∴ $P(k+1)$ is true.

So $P(n)$ is true for all integers $n \geq 2$ by mathematical induction. ✓

Note the first step is to prove $P(2)$ true, not $P(1)$. When attempting to prove $P(k+1)$ is true, often factorising the common factors rather than expanding all of the brackets will lead to a more elegant result. A lot of tedious algebra here, a feature of Maths Extension 2. The HSC markers recommend writing in large, neat print, clearly spreading out your work.

d   RTP: $\forall n > 1, n \in \mathbb{N}, \log_n (n + 1)$ is irrational

Proof by contradiction:

Assume that $\log_n (n + 1)$ is rational.

That is, $\exists p, q \in \mathbb{N}, q \neq 0$ such that $\log_n (n + 1) = \dfrac{p}{q}$.   ✓

Then

$n^{\frac{p}{q}} = (n + 1)$

$n^p = (n + 1)^q$.

If n is even, then $n + 1$ is odd, then n^p is even and $(n + 1)^q$ is odd so they can't be equal.

If n is odd, then $n + 1$ is even, then n^p is odd and $(n + 1)^q$ is even so they can't be equal.   ✓

Contradiction.

Therefore , $\forall n > 1, n \in \mathbb{N}, \log_n (n + 1)$ is irrational.   ✓

> A common proof by contradiction. Learn to set it out correctly. The above method is the most efficient.

Question 15

a i   RTP: If $k + 1$ is divisible by 3 then $k^3 + 1$ is divisible by 3.

Let $k + 1 = 3M$ for some positive integer M.   ✓

$\therefore k = 3M - 1$

$\begin{aligned} \therefore k^3 + 1 &= (3M - 1)^3 + 1 \\ &= (3M)^3 - 3(3M)^2 + 3(3M) - 1 + 1 \\ &= 27M^3 - 27M^2 + 9M \\ &= 3(9M^3 - 9M^2 + 3M), \text{ which is divisible by 3} \end{aligned}$ ✓

OR

Let $k + 1 = 3M$ for some positive integer M.   ✓

Consider the factorisation of sum of 2 cubes:

$\begin{aligned} k^3 + 1 &= (k + 1)(k^2 - k + 1) \\ &= 3M(k^2 - k + 1), \text{ which is divisible by 3.} \end{aligned}$ ✓

> Even though the 'sum and difference of 2 cubes' formula is no longer part of the Maths Advanced course, it has many applications in the Maths Extension 2 course, in proofs like this and with complex numbers.

ii   The contrapositive is: 'If $k^3 + 1$ is not divisible by 3, then $k + 1$ is not divisible by 3.'   ✓

> Swap the if-then phrases, and put 'not' in both of them.

iii The converse is: 'If $k^3 + 1$ is divisible by 3, then $k + 1$ is divisible by 3.' ✓

Swap the if-then phrases.

This converse statement is true.

Proof by contradiction:

Assume $k^3 + 1$ is divisible by 3, but $k + 1$ is not divisible by 3.

So $k + 1 = 3M + 1$ OR $k + 1 = 3M - 1$ for some positive integer M.

If $k + 1 = 3M + 1$, then:

$$k = 3M$$
$$\begin{aligned} k^3 + 1 &= (3M)^3 + 1 \\ &= 27M^3 + 1 \\ &= 3(9M^3) + 1, \text{ which is not divisible by 3} \end{aligned}$$

If $k + 1 = 3M - 1$, then:

$$k = 3M - 2$$
$$\begin{aligned} k^3 + 1 &= (3M - 2)^3 + 1 \\ &= 27M^3 - 3(3M)^2(2) + 3(3M)(2^2) - 8 + 1 \\ &= 27M^3 - 54M^2 + 36M - 7 \\ &= 3(9M^3 - 18M^2 + 12M - 2) - 1, \text{ which is not divisible by 3.} ✓ \end{aligned}$$

Contradiction.

$k + 1$ is divisible by 3.

Therefore, if $k^3 + 1$ is divisible by 3, then $k + 1$ is divisible by 3. ✓

OR

Assume $k^3 + 1$ is divisible by 3, but $k + 1$ is not divisible by 3.

Since $k^3 + 1 = (k + 1)(k^2 - k + 1)$ and $k + 1$ is not divisible by 3, then $(k^2 - k + 1)$ must be divisible by 3.

Now $k^2 - k + 1 = k^2 + 2k + 1 - 3k$
$$= (k + 1)^2 - 3k$$

If $(k + 1)^2 - 3k$ is divisible by 3, then we should be able to factorise 3 from it.

$3k$ is divisible by 3, so $(k + 1)^2$ is divisible by 3 and therefore $k + 1$ is divisible by 3. ✓

Contradiction.

Therefore, if $k^3 + 1$ is divisible by 3, then $k + 1$ is divisible by 3. ✓

This is a difficult proof worth 3 marks, but there are many alternative methods, including the use of the contrapositive. Cases need to be taken to lead to the contradiction. It is also useful to know that one of any 3 consecutive integers is always divisible by 3.

WORKED SOLUTIONS

b i RTP: $\overrightarrow{AC} = \dfrac{n}{m+n}(\underset{\sim}{b} - \underset{\sim}{a})$

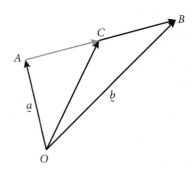

$\dfrac{CB}{AC} = \dfrac{m}{n}.$

So AB is divided into $m + n$ parts.

So $\overrightarrow{AB} = \overrightarrow{AC} + \overrightarrow{CB}$

$\qquad\quad = \dfrac{n}{m+n}\overrightarrow{AB} + \dfrac{m}{m+n}\overrightarrow{AB}.$ ✔

Now,

$\underset{\sim}{a} + \overrightarrow{AB} = \underset{\sim}{b}$

$\qquad \overrightarrow{AB} = \underset{\sim}{b} - \underset{\sim}{a}.$

So

$\overrightarrow{AC} = \overrightarrow{AB}\left(\dfrac{n}{m+n}\right)$

$\qquad = (\underset{\sim}{b} - \underset{\sim}{a})\left(\dfrac{n}{m+n}\right).$ ✔

> It is useful to know how to add and subtract vectors efficiently. The HSC markers reported poor use of vector notation (learn this!). Many students found all of part **b** challenging. Also beware that if given a ratio, such as $\dfrac{CB}{AC} = \dfrac{m}{n}$, it does not mean that $CB = m$ and $AC = n$.

ii $\overrightarrow{OC} = \underset{\sim}{a} + \overrightarrow{AC}$

$\qquad = \underset{\sim}{a} + \left(\dfrac{n}{m+n}\right)(\underset{\sim}{b} - \underset{\sim}{a})$ (from part **i**)

$\qquad = \dfrac{\underset{\sim}{a}m + \underset{\sim}{a}n + \underset{\sim}{b}n - \underset{\sim}{a}n}{m+n}$

$\qquad = \dfrac{\underset{\sim}{a}m + \underset{\sim}{b}n}{m+n}$

$\qquad = \dfrac{m}{m+n}\underset{\sim}{a} + \dfrac{n}{m+n}\underset{\sim}{b}$ ✔

> Adding vectors in the triangle OAC and using part **i** leads to the result quite easily. Some HSC students wrote a page of working for this 1-mark question. During the exam, spend some time looking for the simplest and clearest method, using clues from part **i**.

iii T is the point of intersection of OS and PR.

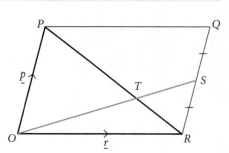

Equation of OS is $\lambda_1\left(\underset{\sim}{r} + \dfrac{1}{2}\underset{\sim}{p}\right)$ [1] ✔

because $\underset{\sim}{p} = \overrightarrow{RQ}$ (opposite sides of a parallelogram).

Equation of PR is $\underset{\sim}{r} + \lambda_2(\underset{\sim}{p} - \underset{\sim}{r})$ [2]

$\qquad \left[\text{or } \underset{\sim}{p} + \lambda_2(\underset{\sim}{r} - \underset{\sim}{p})\right].$

Equating [1] and [2]:

r: $\lambda_1 = 1 - \lambda_2$ [3]

p: $\frac{1}{2}\lambda_1 = \lambda_2$ [4] ✓

Substitute [4] into [3]:

$$\lambda_1 = 1 - \frac{1}{2}\lambda_1$$

$$\frac{3}{2}\lambda_1 = 1$$

$$\lambda_1 = \frac{2}{3}$$

Substitute into [4]:

$$\lambda_2 = \frac{1}{2}\left(\frac{2}{3}\right) = \frac{1}{3}$$

Substitute into [1]:

$$\overrightarrow{OT} = \frac{2}{3}\left(r + \frac{1}{2}p\right)$$

$$= \frac{2}{3}r + \frac{1}{3}p \; ✓$$

Alternative method using similar triangles:

In $\triangle POT$ and $\triangle RST$:

$\angle POT = \angle RST$ (alternate angles, $PO \parallel QR$)

$\angle PTO = \angle RTS$ (vertically opposite angles)

$\therefore \triangle POT \;|||\; \triangle RST$ (equiangular)

$$\frac{SR}{PO} = \frac{SR}{QR} = \frac{1}{2} \quad \text{(opposite sides of a parallelogram)}$$

$$\therefore \frac{SR}{PO} = \frac{ST}{OT} = \frac{1}{2} \quad \begin{array}{l}\text{(matching sides in similar}\\ \text{triangles) ✓}\end{array}$$

$$\overrightarrow{OS} = \overrightarrow{OR} + \overrightarrow{RS}$$

$$= r + \frac{1}{2}p \; ✓$$

As $\dfrac{OT}{ST} = \dfrac{2}{1}$,

$$\frac{OT}{OS} = \frac{OT}{OT + TS} = \frac{2}{2+1} = \frac{2}{3}$$

$$\overrightarrow{OT} = \frac{2}{3}\overrightarrow{OS}$$

$$= \frac{2}{3}\left(r + \frac{1}{2}p\right)$$

$$= \frac{2}{3}r + \frac{1}{3}p. \; ✓$$

The HSC markers advise that drawing large diagrams can help to visualise and solve the problem. Also, practise your vector notation and properties. Note that in this vectors proof, using similar triangle geometry was actually a shorter method.

iv $\overrightarrow{OT} = \dfrac{2}{3}r + \dfrac{1}{3}p$

This is the same situation as part **ii**,

$\overrightarrow{OC} = \dfrac{m}{m+n}a + \dfrac{m}{m+n}b$, where $m = 2$, $n = 1$.

$\therefore \dfrac{PT}{TR} = \dfrac{m}{n} = \dfrac{2}{1}$

$\therefore T$ divides PR in the ratio $2:1$. ✓

OR $\overrightarrow{PT} = \overrightarrow{OT} - p$

$$= \frac{2}{3}r + \frac{1}{3}p - p \quad \text{(from part ii)}$$

$$= \frac{2}{3}r - \frac{2}{3}p$$

$$\overrightarrow{PR} = r - p$$

$$\therefore \overrightarrow{PT} = \frac{2}{3}\overrightarrow{PR}$$

$\therefore T$ divides PR in the ratio $2:1$. ✓

Alternative method using similar triangles:

From part **iii**:

$\dfrac{RT}{PT} = \dfrac{1}{2}$ (matching sides in similar triangles)

$\therefore PT:TR = 2:1$ and T divides PR in the ratio $2:1$. ✓

At first, parts **iii** and **iv** look unrelated to parts **i** and **ii**, but given the hint in part **iv**, the result is easily obtained. This was another 1-mark question that should not require much working.

Question 16

a i $(2m + 4m)\ddot{x} = (4mg - 2mg) - 2kv$ ✓

$$6m\ddot{x} = 2(mg - kv)$$

$$\ddot{x} = \frac{gm - kv}{3m} \checkmark$$

Hence, $\dfrac{dv}{dt} = \dfrac{gm - kv}{3m}$, as required.

> Learn to draw force diagrams correctly. Let the positive direction be downwards, as shown by the dotted arrow in the diagram. Some students neglected to include the tension kv. The net force in this system is equivalent to the two masses working against each other. Note: the net force has a mass of the combined masses. As the larger mass is moving downwards, the smaller mass is moving upwards, resulting in $F_{2m} = 2mg + kv$ and $F_{4m} = 4mg - kv$. As they work against each other, the net force is $6m\ddot{x} = (4mg - kv) - (2mg + kv)$.

ii Given $v < \dfrac{gm}{k}$:

$$\int \frac{dv}{gm - kv} = \int \frac{dt}{3m}$$

$-\dfrac{1}{k} \displaystyle\int \dfrac{-k\, dv}{gm - kv} = \dfrac{1}{3m} \displaystyle\int dt$, adjusting the LHS to make a standard integral. ✓

$-\dfrac{1}{k}\ln(gm - kv) = \dfrac{1}{3m}t + c \qquad v < \dfrac{gm}{k}$, so $kv < gm$ and $gm - kv > 0$

When $t = 0$, $v = 0$:

$$-\frac{1}{k}\ln(gm - 0) = 0 + c$$

$$\therefore c = -\frac{1}{k}\ln(gm)$$

$$-\frac{1}{k}\ln(gm - kv) = \frac{1}{3m}t - \frac{1}{k}\ln(gm)$$

$$\frac{1}{3m}t = -\frac{1}{k}\ln\left(\frac{gm - kv}{gm}\right)$$

$$= \frac{1}{k}\ln\left(\frac{gm}{gm - kv}\right) \checkmark$$

$$e^{\frac{kt}{3m}} = \frac{gm}{gm - kv}$$

$$gm - kv = gm\, e^{-\frac{kt}{3m}}$$

$$kv = gm - gm\, e^{-\frac{kt}{3m}}$$

$$v = \frac{gm}{k}\left(1 - e^{-\frac{kt}{3m}}\right)$$

When $t = \dfrac{3m}{k}\ln 2$,

$$v = \frac{gm}{k}\left(1 - e^{-\ln 2}\right)$$

$$= \frac{gm}{k}\left(1 - \frac{1}{2}\right) \qquad -\ln 2 = \ln 2^{-1}$$

$$= \frac{gm}{2k}, \text{ as required. } \checkmark$$

> This question applies differential equations to a dynamics problem. Be careful with justifying the removal of the ‖ symbols from the logarithm expression after integrating. The HSC markers reported careless student errors in algebra, logarithms and integration, including forgetting to find constants when integrating.
>
> Note that $\ln\left(\dfrac{1}{a}\right) = -\ln a$ and $e^{\ln(a)} = a$, as these are used in the simplification process.

b i $I_n = \int_0^{\frac{\pi}{2}} \sin^{2n+1}(2\theta)\, d\theta$

$I_n = \int_0^{\frac{\pi}{2}} \sin^{2n}(2\theta)\sin(2\theta)\, d\theta$

Using integration by parts:

$u = \sin^{2n} 2\theta,$ $v' = \sin 2\theta$

$u' = 4n \sin^{2n-1} 2\theta \cos 2\theta$ $v = -\frac{1}{2}\cos 2\theta$ ✓

$$I_n = \left[-\frac{1}{2}\cos 2\theta \, \sin^{2n} 2\theta \right]_0^{\frac{\pi}{2}} + 2n \int_0^{\frac{\pi}{2}} \sin^{2n-1}(2\theta)\cos^2(2\theta)\, d\theta$$

$$= [0 - 0] + 2n \int_0^{\frac{\pi}{2}} \sin^{2n-1} 2\theta \, (1 - \sin^2 2\theta)\, d\theta$$

$$= 2n \int_0^{\frac{\pi}{2}} \sin^{2n-1} 2\theta - \sin^{2n+1} 2\theta \, d\theta \;\; ✓$$

$$= 2n \int_0^{\frac{\pi}{2}} \sin^{2(n-1)+1} 2\theta - \sin^{2n+1} 2\theta \, d\theta$$

$$= 2n\left[I_{n-1} - I_n \right]$$

$$I_n = 2nI_{n-1} - 2nI_n$$

$$I_n + 2nI_n = 2nI_{n-1}$$

$$I_n(1 + 2n) = 2nI_{n-1}$$

$$\therefore I_n = \frac{2n}{2n + 1} I_{n-1}, \text{ as required.} \;\; ✓$$

These challenging proofs and integrals made up the very last question of the 2020 exam. Much practice or trial-and-error is required to break these trigonometric expressions apart to obtain the desired result. The hint is usually in the answer, in this case, you need to get I_{n-1}. This is not always easy to do. When setting out a proof, it is important to show your working clearly and not skip steps. For example, the 4 components of an integration by parts should be listed. Choosing the correct u and v' is crucial.

ii $\therefore I_n = \dfrac{2n}{2n+1}I_{n-1}$

$= \dfrac{2n}{2n+1}\left[\dfrac{2(n-1)}{2(n-1)+1}\right]I_{n-2}$

$= \dfrac{2n}{2n+1}\left[\dfrac{2n-2}{2n-1}\right]\left[\dfrac{2(n-2)}{2(n-2)+1}\right]I_{n-3}$

$= \dfrac{2n}{2n+1}\left[\dfrac{2n-2}{2n-1}\right]\left[\dfrac{2n-4}{2n-3}\right]\left[\dfrac{2n-6}{2n-5}\right]I_{n-4}$

$= \dfrac{2n}{2n+1}\left[\dfrac{2n-2}{2n-1}\right]\left[\dfrac{2n-4}{2n-3}\right]\left[\dfrac{2n-6}{2n-5}\right]\cdots\dfrac{2}{3}I_0$ ✓

But $I_0 = \displaystyle\int_0^{\frac{\pi}{2}} \sin 2\theta\, d\theta$

$= \left[-\dfrac{1}{2}\cos 2\theta\right]_0^{\frac{\pi}{2}}$

$= \left[-\dfrac{1}{2}\cos \pi - -\dfrac{1}{2}\cos 0\right]$

$= \left[-\dfrac{1}{2}(-1) + \dfrac{1}{2}(1)\right]$

$= 1$ ✓

$\therefore I_n = \dfrac{2n}{2n+1}\left[\dfrac{2n-2}{2n-1}\right]\left[\dfrac{2n-4}{2n-3}\right]\left[\dfrac{2n-6}{2n-5}\right]\cdots\dfrac{2}{3}$

$= \dfrac{2(n)2(n-1)2(n-2)2(n-3)\ldots 2(1)}{(2n+1)(2n-1)(2n-3)(2n-5)\ldots 3}$

$= \dfrac{2^n(n!)}{(2n+1)(2n-1)(2n-3)(2n-5)\ldots 3}$

To make the denominator $(2n+1)!$, multiply the numerator and denominator by the product of 'even' terms, $2n(2n-2)(2n-4)\ldots 2$.

$I_n = \dfrac{2^n(n!)}{(2n+1)(2n-1)(2n-3)(2n-5)\ldots 3} \times \dfrac{2n(2n-2)(2n-4)(2n-6)\ldots 2}{2n(2n-2)(2n-4)(2n-6)\ldots 2}$

$= \dfrac{2^n(n!)}{(2n+1)!} \times \dfrac{2n[2(n-1)][2(n-2)][2(n-3)]\ldots 2[1]}{1}$

$= \dfrac{2^n(n!)}{(2n+1)!} \times \dfrac{2^n(n!)}{1}$

$I_n = \dfrac{2^{2n}(n!)^2}{(2n+1)!}$ ✓

> This recurrence relation integral is difficult and requires foresight and spotting a pattern. The HSC markers recommend you show the first 3 terms of the sequence to establish the pattern, and the second to last and last terms. Substituting into the expression from part **i** repeatedly, you get close to the required result, noting what is missing and determining how to go further based on this information is necessary. Many students forgot to find the value of I_0.

WORKED SOLUTIONS

iii $J_n = \int_0^1 x^n (1-x)^n \, dx$

Putting $1 - x = \sin^2 \theta$

$x = 1 - \sin^2 \theta$

$dx = -2 \sin \theta \cos \theta \, d\theta$ ✓

When $x = 0$, $\theta = \dfrac{\pi}{2}$ and $x = 1$, $\theta = 0$:

Substituting:

$J_n = \int_{\frac{\pi}{2}}^0 (1 - \sin^2 \theta)^n (\sin^2 \theta)^n (-2 \sin \theta \cos \theta) \, d\theta$

$J_n = -2 \int_{\frac{\pi}{2}}^0 \cos^{2n} \theta \sin^{2n} \theta \sin \theta \cos \theta \, d\theta$

$J_n = 2 \int_0^{\frac{\pi}{2}} \sin^{2n+1} \theta \cos^{2n+1} \theta \, d\theta$ ✓   (swapping the limits of integration)

$\quad = \dfrac{2}{2^{2n+1}} \int_0^{\frac{\pi}{2}} 2^{2n+1} \sin^{2n+1} \theta \cos^{2n+1} \theta \, d\theta$ (making integral ready to convert to $\sin 2\theta$ form)

$\quad = \dfrac{1}{2^{2n}} \int_0^{\frac{\pi}{2}} \sin^{2n+1} 2\theta \, d\theta$ ✓

Using part **ii**: $J_n = \dfrac{1}{2^{2n}} \dfrac{(2^n)^2 (n!)^2}{(2n+1)!}$

$\quad\quad\quad = \dfrac{(n!)^2}{(2n+1)!}$, as required.

Another complex proof requiring a diverse range of mathematical skills, including knowing to take out a factor of $\dfrac{1}{2^{2n}}$. As this is part **iii**, you should expect that it is connected to parts **i** and/or **ii**.

It is necessary to find a connection between the given result for J_n and I_n in part **i**.

The substitution $x = \sin^2 \theta$ or $x = \cos^2 \theta$ also works.

Note that $\sin 2\theta = 2 \sin \theta \cos \theta$. Commit this trigonometric identity to memory as it is used often. It is also on the HSC exam reference sheet should you forget.

iv Now, $f(x) = x(1 - x)$ has a maximum value of $\frac{1}{4}$ when $x = \frac{1}{2}$.

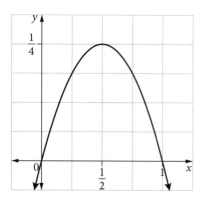

So $x^n(1 - x)^n$ has a maximum value of $\left(\frac{1}{4}\right)^n = \frac{1}{2^{2n}}$. ✓

So the area under the curve $y = x^n(1 - x)^n$ will be less than the area of a rectangle with height $\frac{1}{2^{2n}}$ and width 1.

$$\therefore J_n \le \left[\frac{1}{2^{2n}}x\right]_0^1$$

$$= \frac{1}{2^{2n}}(1 - 0)$$

$$= \frac{1}{2^{2n}}$$

$$\therefore \quad \frac{(n!)^2}{(2n + 1)!} \le \frac{1}{2^{2n}}$$

$$2^{2n}(n!)^2 \le (2n + 1)!$$

$$(2^n n!)^2 \le (2n + 1)!, \text{ as required. } ✓$$

Again, observe the similarity in the part **iii** result and the inequality in the result to be proved in part **iv**.

Noting that $x(1 - x) \le \frac{1}{4}$, you can then obtain an expression for $x(1 - x)$ and finally $x^n(1 - x)^n$ which can be followed up by using J_n and the answer from part **iii** to get the required result.

This question requires you to use logic and explain your reasoning. There are also many alternative solutions to this question, some more complex and time-consuming than others. Choose the simplest and most efficient method.

9780170459273

HSC exam reference sheet

Mathematics Advanced, Extension 1 and Extension 2

© NSW Education Standards Authority

Note: Unlike the actual HSC exam reference sheet, this sheet indicates which formulas are Mathematics Extension 1 and 2.

Measurement

Length

$$l = \frac{\theta}{360} \times 2\pi r$$

Area

$$A = \frac{\theta}{360} \times \pi r^2$$

$$A = \frac{h}{2}(a + b)$$

Surface area

$$A = 2\pi r^2 + 2\pi rh$$

$$A = 4\pi r^2$$

Volume

$$V = \frac{1}{3}Ah$$

$$V = \frac{4}{3}\pi r^3$$

Functions

$$x = \frac{-b \pm \sqrt{b^2 - 4ac}}{2a}$$

For $ax^3 + bx^2 + cx + d = 0$:*   *EXT1

$$\alpha + \beta + \gamma = -\frac{b}{a}$$

$$\alpha\beta + \alpha\gamma + \beta\gamma = \frac{c}{a}$$

$$\text{and } \alpha\beta\gamma = -\frac{d}{a}$$

Relations

$$(x - h)^2 + (y - k)^2 = r^2$$

Financial Mathematics

$$A = P(1 + r)^n$$

Sequences and series

$$T_n = a + (n - 1)d$$

$$S_n = \frac{n}{2}\left[2a + (n - 1)d\right] = \frac{n}{2}(a + l)$$

$$T_n = ar^{n-1}$$

$$S_n = \frac{a(1 - r^n)}{1 - r} = \frac{a(r^n - 1)}{r - 1}, r \neq 1$$

$$S = \frac{a}{1 - r}, |r| < 1$$

Logarithmic and Exponential Functions

$$\log_a a^x = x = a^{\log_a x}$$

$$\log_a x = \frac{\log_b x}{\log_b a}$$

$$a^x = e^{x \ln a}$$

Trigonometric Functions

$$\sin A = \frac{\text{opp}}{\text{hyp}}, \quad \cos A = \frac{\text{adj}}{\text{hyp}}, \quad \tan A = \frac{\text{opp}}{\text{adj}}$$

$$A = \frac{1}{2}ab\sin C$$

$$\frac{a}{\sin A} = \frac{b}{\sin B} = \frac{c}{\sin C}$$

$$c^2 = a^2 + b^2 - 2ab\cos C$$

$$\cos C = \frac{a^2 + b^2 - c^2}{2ab}$$

$$l = r\theta$$

$$A = \frac{1}{2}r^2\theta$$

Trigonometric identities

$$\sec A = \frac{1}{\cos A}, \quad \cos A \neq 0$$

$$\csc A = \frac{1}{\sin A}, \quad \sin A \neq 0$$

$$\cot A = \frac{\cos A}{\sin A}, \quad \sin A \neq 0$$

$$\cos^2 x + \sin^2 x = 1$$

Compound angles*

$$\sin(A + B) = \sin A \cos B + \cos A \sin B$$

$$\cos(A + B) = \cos A \cos B - \sin A \sin B$$

$$\tan(A + B) = \frac{\tan A + \tan B}{1 - \tan A \tan B}$$

If $t = \tan\dfrac{A}{2}$, then $\sin A = \dfrac{2t}{1 + t^2}$

$$\cos A = \frac{1 - t^2}{1 + t^2}$$

$$\tan A = \frac{2t}{1 - t^2}$$

$$\cos A \cos B = \frac{1}{2}\big[\cos(A - B) + \cos(A + B)\big]$$

$$\sin A \sin B = \frac{1}{2}\big[\cos(A - B) - \cos(A + B)\big]$$

$$\sin A \cos B = \frac{1}{2}\big[\sin(A + B) + \sin(A - B)\big]$$

$$\cos A \sin B = \frac{1}{2}\big[\sin(A + B) - \sin(A - B)\big]$$

$$\sin^2 nx = \frac{1}{2}(1 - \cos 2nx)$$

$$\cos^2 nx = \frac{1}{2}(1 + \cos 2nx)$$

Statistical Analysis

$$z = \frac{x - \mu}{\sigma}$$

An outlier is a score less than $Q_1 - 1.5 \times \text{IQR}$ or more than $Q_3 + 1.5 \times \text{IQR}$

Normal distribution

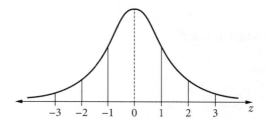

- approximately 68% of scores have z-scores between -1 and 1

- approximately 95% of scores have z-scores between -2 and 2

- approximately 99.7% of scores have z-scores between -3 and 3

Discrete random variables

$$E(X) = \mu$$

$$\text{Var}(X) = E\big[(X - \mu)^2\big] = E(X^2) - \mu^2$$

Probability

$$P(A \cap B) = P(A)P(B)$$

$$P(A \cup B) = P(A) + P(B) - P(A \cap B)$$

$$P(A|B) = \frac{P(A \cap B)}{P(B)}, \quad P(B) \neq 0$$

Continuous random variables

$$P(X \leq r) = \int_a^r f(x)\,dx$$

$$P(a < X < b) = \int_a^b f(x)\,dx$$

Binomial distribution*

$$P(X = r) = {}^nC_r\, p^r (1 - p)^{n-r}$$

$$X \sim \text{Bin}(n, p)$$
$$\Rightarrow P(X = x)$$
$$= \binom{n}{x} p^x (1 - p)^{n-x}, \quad x = 0, 1, \ldots, n$$

$$E(X) = np$$

$$\text{Var}(X) = np(1 - p)$$

*EXT1

Differential Calculus

Function	Derivative
$y = f(x)^n$	$\dfrac{dy}{dx} = nf'(x)[f(x)]^{n-1}$
$y = uv$	$\dfrac{dy}{dx} = u\dfrac{dv}{dx} + v\dfrac{du}{dx}$
$y = g(u)$ where $u = f(x)$	$\dfrac{dy}{dx} = \dfrac{dy}{du} \times \dfrac{du}{dx}$
$y = \dfrac{u}{v}$	$\dfrac{dy}{dx} = \dfrac{v\dfrac{du}{dx} - u\dfrac{dv}{dx}}{v^2}$
$y = \sin f(x)$	$\dfrac{dy}{dx} = f'(x)\cos f(x)$
$y = \cos f(x)$	$\dfrac{dy}{dx} = -f'(x)\sin f(x)$
$y = \tan f(x)$	$\dfrac{dy}{dx} = f'(x)\sec^2 f(x)$
$y = e^{f(x)}$	$\dfrac{dy}{dx} = f'(x)e^{f(x)}$
$y = \ln f(x)$	$\dfrac{dy}{dx} = \dfrac{f'(x)}{f(x)}$
$y = a^{f(x)}$	$\dfrac{dy}{dx} = (\ln a)f'(x)a^{f(x)}$
$y = \log_a f(x)$	$\dfrac{dy}{dx} = \dfrac{f'(x)}{(\ln a)f(x)}$
$y = \sin^{-1} f(x)$	$\dfrac{dy}{dx} = \dfrac{f'(x)}{\sqrt{1 - [f(x)]^2}}$ *
$y = \cos^{-1} f(x)$	$\dfrac{dy}{dx} = -\dfrac{f'(x)}{\sqrt{1 - [f(x)]^2}}$ *
$y = \tan^{-1} f(x)$	$\dfrac{dy}{dx} = \dfrac{f'(x)}{1 + [f(x)]^2}$ *

Integral Calculus

$$\int f'(x)[f(x)]^n \, dx = \frac{1}{n+1}[f(x)]^{n+1} + c$$
$$\text{where } n \neq -1$$

$$\int f'(x)\sin f(x) \, dx = -\cos f(x) + c$$

$$\int f'(x)\cos f(x) \, dx = \sin f(x) + c$$

$$\int f'(x)\sec^2 f(x) \, dx = \tan f(x) + c$$

$$\int f'(x)e^{f(x)} \, dx = e^{f(x)} + c$$

$$\int \frac{f'(x)}{f(x)} \, dx = \ln|f(x)| + c$$

$$\int f'(x)a^{f(x)} \, dx = \frac{a^{f(x)}}{\ln a} + c$$

$$\int \frac{f'(x)}{\sqrt{a^2 - [f(x)]^2}} \, dx = \sin^{-1}\frac{f(x)}{a} + c \; *$$

$$\int \frac{f'(x)}{a^2 + [f(x)]^2} \, dx = \frac{1}{a}\tan^{-1}\frac{f(x)}{a} + c \; *$$

$$\int u\frac{dv}{dx} \, dx = uv - \int v\frac{du}{dx} \, dx \; **$$

$$\int_a^b f(x) \, dx$$
$$\approx \frac{b-a}{2n}\{f(a) + f(b) + 2[f(x_1) + \cdots + f(x_{n-1})]\}$$
$$\text{where } a = x_0 \text{ and } b = x_n$$

*EXT1, **EXT2

9780170459273

Combinatorics*

$$^nP_r = \frac{n!}{(n-r)!}$$

$$\binom{n}{r} = {^nC_r} = \frac{n!}{r!(n-r)!}$$

$$(x+a)^n = x^n + \binom{n}{1}x^{n-1}a + \cdots + \binom{n}{r}x^{n-r}a^r + \cdots + a^n$$

Vectors*

$$\left|\underset{\sim}{u}\right| = \left|x\underset{\sim}{i} + y\underset{\sim}{j}\right| = \sqrt{x^2 + y^2}$$

$$\underset{\sim}{u} \cdot \underset{\sim}{v} = \left|\underset{\sim}{u}\right|\left|\underset{\sim}{v}\right|\cos\theta = x_1 x_2 + y_1 y_2,$$
where $\underset{\sim}{u} = x_1\underset{\sim}{i} + y_1\underset{\sim}{j}$
and $\underset{\sim}{v} = x_2\underset{\sim}{i} + y_2\underset{\sim}{j}$

$$\underset{\sim}{r} = \underset{\sim}{a} + \lambda\underset{\sim}{b}**$$

Complex Numbers**

$$z = a + ib = r(\cos\theta + i\sin\theta)$$
$$= re^{i\theta}$$

$$\left[r(\cos\theta + i\sin\theta)\right]^n = r^n(\cos n\theta + i\sin n\theta)$$
$$= r^n e^{in\theta}$$

Mechanics**

$$\frac{d^2x}{dt^2} = \frac{dv}{dt} = v\frac{dv}{dx} = \frac{d}{dx}\left(\frac{1}{2}v^2\right)$$

$$x = a\cos(nt + \alpha) + c$$

$$x = a\sin(nt + \alpha) + c$$

$$\ddot{x} = -n^2(x - c)$$

*EXT1, **EXT2